Wilder Lives

WILDER LIVES
Humans and Our Environments

DUNCAN BROWN

 UNIVERSITY OF KWAZULU-NATAL PRESS

Published in 2019 by University of KwaZulu-Natal Press
Private Bag X01
Scottsville, 3209
Pietermaritzburg
South Africa
Email: books@ukzn.ac.za
Website: www.ukznpress.co.za

ISBN: 978 1 86914 421 0
e-ISBN: 978 1 86914 422 7

Editor: Sally Hines
Proofreader: Catherine Munro
Typesetter: Patricia Comrie
Indexer: Judith Shier
Cover design: Marise Bauer, M Design
Cover artwork: Wayne Robertson (Hipebeast Photography)

For the author and perfecter of wild love.

And for Tracey and Michael, who always just got it.

Virtually all contemporary people are cultivated stock, but we can stray back into the woods.

— Gary Snyder, *The Practice of the Wild*

When despair for the world grows in me
and I wake in the night at the least sound
in fear of what my life and my children's lives may be,
I go and lie down where the wood drake
rests in his beauty on the water, and the great heron feeds.
I come into the peace of wild things
who do not tax their lives with forethought
of grief. I come into the presence of still water.
And I feel above me the day-blind stars
waiting with their light. For a time
I rest in the grace of the world, and am free.

(Wendell Berry, 'The Peace of Wild Things')

Contents

Acknowledgements

Many people have assisted me in the researching and writing of this book, whether through conversations, lending me material, sending me links, hosting me, commenting on drafts or sections, or just keeping me motivated: Stephen Boshoff; Craig Thom; Julia Martin; Wendy Woodward; Antjie Krog; Alan and Annabelle Hobson; Wolf Avni; Michael Chapman; Tom Sutcliffe; Kobus van der Merwe; Michael Wessels; Hermann Wittenberg; Knut Nustad; Ted Chamberlin; Mike Davies-Coleman; Vivienne Lawack; Michael Brown; Ken Brown; and Judy Brown.

I am grateful to the National Research Foundation (NRF) and the Andrew W. Mellon Foundation for funding that assisted me in conducting this research. Opinions expressed and research findings are my own, and are not necessarily to be attributed to the NRF or the Mellon Foundation.

I am also grateful to the Stellenbosch Institute for Advanced Study (STIAS) for two fellowships that enabled me to give full attention to this research. I must also thank the fellows of STIAS who discussed this project with me, and who commented on my seminar presentations.

I am indebted to my deputy deans, Uma Dhupelia-Mesthrie, Lindsay Clowes and Steward van Wyk, for their generosity in acting as dean at various times so that I could undertake fellowships at STIAS, and other research trips related to this project.

Stephen Boshoff and Antjie Krog read and commented on the first draft of this manuscript, and offered me sage advice on what to revise, cut, expand or rewrite. Their input was invaluable. Any infelicities that remain are my own.

The feedback from the anonymous readers appointed by the University of KwaZulu-Natal Press proved invaluable, and has improved the final version of the book.

Special thanks are due to Sally Hines for her meticulous and sensitive editing, and for her enthusiasm for this project. Thank you to Catherine Munro for her excellent proofreading. I am very grateful to UKZN Press and its staff for bringing the book to publication.

Tracey Brown is my first and best editor, and companion, reading the pages as they came off the printer, and also walking with me every step of the way through this book. Without her, this book would have been much the poorer.

1

Into the Wild

Wilderness may temporarily dwindle, but wildness won't go away.

— Gary Snyder, *The Practice of the Wild*

A couple of years ago I had lunch at a restaurant called Oep ve Koep (Open for Shopping/Business) in the small town of Paternoster up the West Coast of South Africa, run by an enterprising young chef by the name of Kobus van der Merwe. The restaurant's culinary philosophy was to celebrate regional recipes, and local, seasonal produce, including seaweeds. In line with global trends in this regard, it showcased foods that are foraged, and vegetables and herbs included wild rosemary, dune spinach, indigenous sage and sea lettuce, which were collected rather than cultivated. Proteins included locally hunted venison, and the prolifically present angelfish, caught off Saldanha Bay. These were all 'indigenous', 'wild' foodstuffs. The kabeljou was, however, farmed, sourced from a fish farm in Paternoster, as this accorded better with the restaurant's 'sustainable use' philosophy than the harvesting of rapidly declining ocean or estuarine stocks. The mussels, while locally collected, were of the European 'invasive' species (Mediterranean mussels), according to Van der Merwe, as harvesting these enabled the 'indigenous' stocks to regrow. 'Wild indigenes', 'farmed indigenes' and 'wild aliens' all found a place on the menu in the name of local cuisine and sustainability.

Particularly with the rapid growth in global environmental awareness, as well perhaps more cynically of large corporations seeking to conceal destructive practices by 'renaming' them, the meanings, values and ethics associated with terms such as 'wild' and 'farmed', alongside those that modify or modulate them – feral, domesticated, tame, cultivated, to name but a few – are shifting rapidly. To illustrate from the marketing of fish twenty years ago, most gastronomes would universally have opted for 'wild' over 'farmed' salmon, in terms of the perceived superiority in texture and flavour of the former. In South Africa and elsewhere, with increased awareness of pressures on fish stocks, the pendulum is swinging back, with many restaurants or upmarket supermarkets announcing proudly that the fish they sell is 'sustainably farmed'. Fish, such as hake, which may be advertised as 'wild caught', is usually qualified by the rider 'from sustainable stocks', and is identified as such by its listing under the Southern African Sustainable Seafood Initiative (SASSI). When I started fly fishing in what is now KwaZulu-Natal in the 1970s, the universal wisdom was that the only trout really worth eating were wild fish caught in rivers. Stocked stillwater trout were judged to taste 'muddy', and hatchery fish were (correctly in my opinion) said to have a livery taste, deriving from the pellets on which they were fed. In the Western Cape, that logic does not hold true now, if it ever did. (Wild) river trout in this region tend to be rather insipid in taste, largely because the water is very acidic. In contrast, hatchery trout are raised in cold, clear water in the Western Cape winters (the primary rain fall period in other regions of South Africa is summer, during which temperatures are not ideal for growing trout); they are fed on a diet of pellets made from sustainably harvested marine baitfish, and are delicious to eat, approximating in taste, I imagine, to the highly prized sea-run (anadromous) trout of Europe and the Americas. The flesh has the distinctively pinky-orange colour of wild fish that have fed on crustaceans, through 'natural' colourants that are added to the food. Farmed trout in South Africa have, in fact, been awarded SASSI green status as environmentally sustainable. Their taste of 'wildness' is, of course, carefully 'cultivated'.

* * *

What would the world be, once bereft
Of wet and of wildness? Let them be left,
O let them be left, wildness and wet;
Long live the weeds and the wilderness yet.
            (*Gerard Manley Hopkins, 'Inversnaid'*)

* * *

The term 'wild' is everywhere. In the titles of films, television series, music albums and books; in the names of foodstuffs, plants and animals; in place names, brand names, the names of restaurants, the names of companies; in metaphors and idioms; and in describing sources, origins or climatic conditions. *Wild at Heart*, *A Year in the Wild*, wild animals, *Wilder Mind*, go wild, wildlife, wild geese, *The Wild*, wild caught, *The Wild One*, *Wildside*, wild rice, *The Wild Geese*, the call of the wild, 'Near Wild Heaven', wild abandon, wild horses, Wildtrak, wilderness, *Wild Religion*, the Wilderness, wild weather, the Wild Coast, Wild Waters, wild flowers, wild harvested, wild card, wild boars, Wild Cards, the Wild Fig, wild journeys, wild fires, the World Wildlife Foundation, wild country, 'Wild Thing', wild ones, *Wild Planet*, wet 'n wild, *Into the Wild*, Wild Turkey, wild-eyed, wild oats, WildFly, *Wild Law*, wild dogs, wild guesses, wild assumptions, wild conclusions, wild adventures, Wild Bill, Wild West, back to the wild, wild rocket, wild winds, wild goose chase. The term is used so widely and frequently, and often without any qualification, that one would imagine that its meaning were uncontroversial. As with other terms that we use so frequently though – history, nation, culture, to name some of the most obvious – that is far from the case.

A quick read through the usages of 'wild' listed above indicates that there is a great deal of ambiguity, even contradiction, in the meanings associated with the word. But the distinction between the 'wild' and the 'farmed', the 'wild' and the 'civilised', the 'wild' and the 'tame', the 'wild' and the 'domesticated,' even the 'wild' and the 'reasoned', remains fundamental to how we understand and order the world in which we live. The distinction is sometimes understood as that between the 'natural' and the 'unnatural', behind which, of course, lurks the problematic distinction between 'nature' and 'culture'. Like just about all such definitions, that which seeks to define wildness in terms of what it is not (typically the absence of humans and our influence) dissolves under even the slightest interrogation. The philosopher Holmes Rolston points out in this respect:

Humans depend on airflow, water cycles, sunshine, photo-synthesis, nitrogen fixation, decomposition, bacteria, fungi, the ozone layer, food chains, insect pollination, soils, earthworms, climates, oceans, and genetic materials. An ecology always lies in the background of culture, natural givens that support everything else. Some sort of inclusive environmental fitness is required of even the most advanced culture.[1]

So the larger question I want to explore in this book is whether we can find ways of rethinking 'wildness' that are more enabling in understanding the complexities of human interactions with, and responsibilities towards, the biological systems and processes on which all life depends, including ways that may decentre the human? If the Earth is indeed 4.5 billion years old, as scientists currently tell us, recognisably human life has only been around since the last ice age (about 12 000 to 13 000 years ago), and as a species we have single-handedly destroyed our planet's ecosystems in the short space of a few hundred years of industrialisation and rapacious modernity, then we urgently need to rethink and redefine our identities and behaviours as a species. Can 'thinking wild' help? Can it provide different ways of seeing, engaging, being? Can we think of 'wildness' as something that may exist in gradations, or as quality rather than absolute value, and as something that has important ethical dimensions? Can it lead us to a 'world view locating humans in a satisfactory residence on this historic and storied Earth', to use Rolston's suggestive formulation?[2]

To think wild would mean a profound humility on our part, both as individuals and as a species. It would mean acknowledging the mess we have made; a commitment to less hubristic action or no action at all if appropriate; a profound awareness that other species do not necessarily need us, and would likely be better off without us, but that *we* need *them*; that with all of our scientific knowledge we are still woefully ignorant about the interdependencies of life forms; that our desires, wants and needs are not paramount; and that it is possible, and desirable, to live more lightly on the earth.

Thinking wild may also mean to look with different eyes, to expand our hearing, to be more fully present in the body and in the world. We may begin to observe more fully:

Closer scrutiny shows the bird streaking between office blocks above the heads of the hipsters and drug dealers of Long Street to be, not a feral pigeon, but a peregrine falcon.

The oppressive heat and strong winds promise a mammoth cold front. You can almost feel the atmospheric pressure dropping in your chest cavity.

The air is frigid, the landscape gleaming white with snow. By the roadside children are throwing snowballs and parents are stamping their feet in the cold and sipping coffee from thermos mugs. My mind is somewhere between wonder at the magical winter transformation of these scraggly brown fields into blankets of purest white, and the sure knowledge that the icy snowmelt to follow will recharge the rivers and their spawning trout.

The frenetic barking of the dogs the previous night is finally explained by the tracks of a genet in our back garden.

Traversing the sweep of calm, blue, windless seascape, the eye is caught by a flash of white water – signalling gannet, whale, porpoise, shark or large game fish.

More reliable than the calendar in marking the onset of spring is the first sighting of a yellow-billed kite back from its winter migration.

This boerboel does not seem keen to welcome me onto his turf, and I doubt that he weighs much less than I do. An altercation between the two of us is going to be decidedly one-sided. His owners have done little or nothing to socialise him with other dogs or humans. I bend down so that I am not a threatening presence towering over him, and let him approach me, which he now does with little aggression.

I am initially dumbfounded to see whole beds of opened shells with no mussels inside. And then I recall the previous month's red tide that turned the entire bay a deep crimson.

Instinct suggests that what I have just heard is a fish eagle, but reason reminds me that I am in the parking lot of Somerset Mall. Against reason I squint up into the sky to spot the pair of raptors riding the thermals high above the town.

This does not only bring personal pleasure. It enables us to monitor, at least on an informal basis, what is still there in apparently healthy populations. A friend of my son's commented that on her drive to our house she and her father had seen a beautiful hawk. When I pressed her for details, it became clear that it had been a black-shouldered kite. 'Hawk' versus 'black-shouldered kite' is an important distinction. In the 1970s and early 1980s, as a schoolboy I was fascinated with raptors, and desperate to take up the sport of falconry. I scoured the sky and the vegetation on every opportunity, and recall that – with the effects of poisoning and shooting by farmers – pretty much all one routinely saw were steppe buzzards, yellow-billed kites, various kestrels and black-shouldered kites. Certainly few, if any, of the hawks – goshawks, sparrowhawks – or eagles, harriers, or the other buzzards were in evidence, except in remote areas. Thanks to dedicated conservation and awareness campaigns, populations across the board have improved, and from my suburban garden I regularly see goshawks, sparrowhawks, falcons, gymnogenes, various buzzards, and at least two types of eagle, not to mention owls. If any bird of prey meets your category of 'hawk', you will probably have been oblivious to the cataclysmic decline in numbers in the past, and to their heartening resurgence (unless you noted that there were fewer grey ones than before . . .).

The ability to read landscape, or sea, is critical in this regard. My brother has been a merchant seaman all of his working life, and has spent more time gazing at seascapes than most. When he first went to sea in the mid-seventies, he said that as a result of whaling activities, the sight of a whale was so rare – occurring once or twice a month at most – that they were required to keep a log of sightings for conservation authorities. With restrictions on whaling activities, whale populations slowly started to increase, and I recall on family holidays at our cottage in the Eastern Cape, he would occasionally say, after gazing out to sea for what seemed like ages, 'There's a whale out there.' Not knowing the signs, the rest of us would squint hopefully for evidence of something among the mass of white horses. Whale populations globally have now increased to the point that my brother sees hundreds of whales every day, and from our beach cottage window we even see them breaching a hundred metres or so offshore. And with practice, our sea eyes have improved sufficiently that we can tell the difference between the spray put up by a diving gannet and a whale blow, and can spot whales, dolphins, seals, baitfish and game fish even among the white horses. We are learning to see wild.

2

Rewilding

I went to the woods because I wished to live deliberately, to front only the essential facts of life, and see if I could not learn what it had to teach, and not, when I came to die, discover that I had not lived. I did not wish to live what was not life, living is so dear, nor did I wish to practise resignation, unless it was quite necessary. I wanted to live deep and suck out all the marrow of life, to live so sturdily and Spartan-like as to put to rout all that was not life, to cut a broad swath and shave close, to drive life into a corner, and reduce it to its lowest terms, and, if it proved to be mean, why then to get to the whole and genuine meanness of it, and publish its meanness to the world; or if it were sublime, to know it by experience, and be able to give a true account of it in my next excursion.

— Henry David Thoreau, *Walden*

In this chapter, I begin by looking at some understandings of wildness that seem to me problematic and limiting, and then explore what I would see as more productive, suggestive and, I hope, energising ways of thinking wildness. In doing so, I draw especially on Gary Snyder's ideas about 'wildness', and George Monbiot's notion of 'rewilding'.

The terms 'wildness' and 'wilderness' are very closely associated, whether in the language of conservation and environmental management, or in the tourist literature that promotes leisure destinations such as game lodges, nature reserves or fishing camps: the 'wildness' of a place is what guarantees its status as 'wilderness', and these are qualities that draw visitors, especially those whose working lives are spent in cities. This sense of wildness is captured in a feature in *Getaway* magazine (the magazine's title is itself telling) on the 'nine wildest experiences close to home':

> Scott Ramsay has been to some of the most untouched regions on the planet. In all his years of exploring, he counts his visits to the national parks and reserves of Southern Africa among the wildest. 'It's a privilege living in this part of the world,' he says, 'because all of the following experiences are within a day or two of travel from the major cities.'[1]

This sense of untouched 'wildness' resonates strongly with the definition of 'wilderness' set out in the 1964 United States Wilderness Act: 'An area where the earth and its community of life are untrammeled by man, where man himself is a visitor who does not remain.'[2] The scientist Peter B. Landres and his co-authors, who quote this definition, argue that '[s]ynonyms for untrammeled include unimpeded, unhampered, uncontrolled, self-willed and free', and that 'the word "wildness" strongly connotes this sense of an area free from human control or manipulation'.[3] In this understanding, 'wildness' exists only where humans are absent, which renders it questionable as a concept in a world in which human influence is omnipresent (our current era is frequently described in this regard as the Anthropocene), in particular through changes we have wrought to weather patterns, and in which human non-engagement in environmental issues is no longer an ethical or environmental option (if it ever was). Arguments that can only understand human involvement as 'unnatural', and so in a real sense separate the 'human' from the 'biological', seem to me particularly unhelpful.[4] There is, actually, a deep ambiguity in these conceptions of wildness: they can be read as fundamentally anthropocentric (wildness is that from which humans voluntarily absent themselves); or profoundly biocentric (wildness cannot admit of human presence). Neither position seems to offer many possibilities.

Of course, the commercial operations advertising the 'wildness' of reserves, which are in reality carefully 'managed', are involved in a double bluff, more of which in the next chapter on wildness and conservation. But as Landres and his co-authors point out, 'naturalness' and 'wildness' are two concepts that 'strongly influence nearly all wilderness management', and they point to the 'dilemma and irony that arise when wilderness managers contemplate manipulating the environment to restore naturalness at the risk of reducing wildness'.[5] It is a dilemma, contradiction or paradox (depending on one's position) at the very heart of the assumptions and practices of 'managing wilderness', or more broadly of 'nature conservation'.

Landres and his co-authors point out that 'deciding when to take action [is] the central dilemma in wilderness management',[6] and that contemplating intervention necessarily involves deciding between the values of wildness and naturalness:

> In these situations, where human-caused impacts have caused wholesale changes to the wilderness environment, should the wildness of present-day wilderness be compromised to restore naturalness? In other words, should an undesirable means, such as manipulation of wilderness, be used to achieve a desirable end, such as restoration of natural conditions in wilderness?[7]

They identify some searching questions to be asked in such cases:

> Does manipulation compromise the very values that are protected and preserved in wilderness? Is there sufficient technical knowledge to use large-scale manipulation to restore wilderness landscapes? What are the consequences and risks

of taking action versus not taking action? Does the public sufficiently trust the agency to allow such large-scale actions? Does the desire to restore the ecological value of naturalness outweigh the social value of wildness? How much trammeling is necessary and tolerable in wilderness? Is it appropriate to even define a target for desired future ecological conditions in wilderness? Must we accept, albeit reluctantly, the human 'gardenification' of wilderness [. . .]?[8]

They refer to G.A. Fine's identification of 'three overarching philosophical views of the relationship between nature and culture that have predominated over the course of human history'. The first is the 'utilitarian' view, according to which the natural world is a 'storehouse of goods that can meet human needs'. In the second, the 'preservation' perspective, nature can exist '*in spite of* culture', and the protection of nature from human influence is prized. The third is the 'organic' perspective, which is 'both the oldest and newest orientation towards nature', in which the 'the natural world and human world are integrated and even inseparable'.[9] Where we stand in our philosophical understanding of nature and culture will affect how and whether we contemplate environmental intervention, and the ambiguity of wilderness policies in this respect exacerbates the paradoxes and contradictions involved:

The dilemma we face – whether to err on the side of wildness by stressing the nature/culture dichotomy, or to err on the side of naturalness by restoring nature whenever possible – is rooted in the ongoing ambiguity of a wilderness policy and other environmental policies that are rooted both in the preservationist and organic views of nature and culture. Where we fall on the spectrum from dichotomy to convergence is often rooted in our view of risk and uncertainty: Do we dare trust science? Do we dare not?[10]

Landres and his co-authors conclude that 'with our increased knowledge of regional-scale human impacts, coupled with our desire to restore areas known to be degraded, "doing the right thing" is no longer a simple path because it is based on a philosophical choice between wildness and naturalness'.[11] Monbiot, whose work I will discuss a little later, would probably argue, in contrast, that since human actions have already 'dewilded' almost all environments, human intervention to restore species and hence naturalness, for example, would be the starting point for 'rewilding' in which the environment was then left to 'get on with it'.[12]

* * *

Fierce winds and black thunderheads stacking up suggest that I have little time to get off the water before a storm of biblical proportions unleashes itself. I paddle the float tube to the nearest bank, hastily dismantle my rods, and fling everything into the bag, thinking only of refuge from the exposure of these highlands, almost on the roof of the world. The walk back to the car from the far dam at Highmoor is arduous, laden down as I am with gear. The first fat raindrops render the path underfoot greasy and increasingly treacherous, and the air is heavy with menace and static. Eyes on the ground, and thinking only of a warm bath and a whisky, I track the path into a shallow gully. A warning bark brings me to a swift halt. A lone male baboon sizes me up from the path ascending the other side of the gully. He is going nowhere. I am suddenly aware that I am utterly alone and vulnerable in this meeting of species. Casting dignity to the winds I give the baboon best, and abandon the path for a hasty, circuitous, humbling retreat through brambles and briars.

* * *

The environmental geographer Ben Ridder offers a way of thinking about naturalness and wildness in less absolute terms by suggesting we recast the debate as '"protecting biodiversity" versus "respect for nature's autonomy"'.[13] It is a reformulation that I find agreeable. He also points out that there are 'many human artifacts, activities, and attributes that are considered natural, or relatively so', which 'include natural foods and medicines, natural birth of children, and natural mental or physical abilities'.[14] Ridder argues that '[n]aturalness of this kind is highly prized within contemporary Western societies', but it is compromised by 'the involvement of increasing numbers of people, the complexity of the technology used, and the increasing distance between the immediate purpose of the task at hand and the ultimate instrumental objectives of those involved'.[15] In the current desire in Western-influenced consumer cultures for craft beers, artisanal breads, organic foods, heirloom vegetables and slow food, we see a desire to close the human and technological gap between production and consumption (as well, perhaps, as a greater awareness about nutrition and health).

Protecting 'naturalness', or in Ridder's conception, 'biodiversity', almost always, however, involves a historical choice. Landscapes and environments are in a state of perpetual change, and this was the case even when hunting and gathering were the sole means of human food supply. The crucial question is, then, which historical landscape do you choose to restore? What period in time represents the 'authentic' landscape? Monbiot points out that most often as humans, and even as scientists, we assume that we should restore environments to what they were when we were children, and that this fallacious assumption is repeated generation after generation.[16] But even if we acknowledge that as children we inherited a world already severely depleted, how far back do we go, in looking for a model: 100 years, 1 000, 10 000? The answer to that – which is at base asking what a properly indigenous landscape looked like – is something frequently presented by scientists as historical fact, in terms of which species can then be removed from the environment (the debate about trout in South Africa is a case in point). But a reading of scientific literature in various sub-disciplines reveals radically discrepant assumptions about time frames, with some plant scientists suggesting a measure of about 100 years, and others working on animals positing that 5 000 years is not sufficient a span of time.[17]

Ridder uses the example of Australia to show up the complexities and potential fallacies of such arguments. He points out that in that country 'the generally accepted date against which the naturalness of vegetation condition is assessed is 1750'.[18] While acknowledging that European settlement resulted in dramatic changes to the environment, he argues that 'there is general consensus that Aboriginal hunting and the use of fire also gave rise to considerable change in the Australian environment, commencing with their arrival about 56 000 years ago'. In particular, 'the introduction of dogs to the continent 4000–5000 years ago, most probably by fishermen from the northeast of Australia, led to numerous extinctions, including the Tasmanian tiger (*Thylacinus cynocephalus*) and Tasmanian devil (*Sarcophilus harrisii*)', those two species being found at the time of European settlement only on Tasmania, which had been separated off from the mainland by rising sea levels between 8 000 and 10 000 years ago.[19] Ridder cites the extreme example of T.F. Flannery, who in 1994 argued for a restoration of the Australian environment to what it was '60 000 years ago prior to the arrival of the ancestors of the Aborigines', which included reintroducing the Tasmanian devil and introducing the Komodo dragon (*Varanus komodoensis*), which is '"in ecological terms, the closest living species to any of Australia's lost reptilian carnivores"'.[20]

The problem with these arguments about restoration and indigeneity is that 'a particular historical period [is] accorded priority'. But given that the climatic and geomorphological conditions of Australia today are very different from what they were 500 years ago, let alone 60 000 years ago, such restoration efforts will 'demand particular ongoing management actions, the need for which is inconsistent with "natural authenticity"'.[21] In so doing, as humans we make a conservation rod for our own backs, in that we end up having to intervene more and more to protect something that we define as 'natural'.

Ridder then makes the point, fundamental to what I am exploring in this book, 'that many human impacts do not cause nonhuman behavior to become any less spontaneous [. . .], and in consequence, the land is no less wild'.[22] This suggests a conception of wildness that can indeed admit of human presence. I live in what the South African novelist Michiel Heyns describes as the 'darkest suburbia of Somerset West', but my garden is visited by or home to a wide range of birds, including raptors, which exhibit the same wariness of humans as those I encounter elsewhere. When I go for my evening run through Radloff Park, an urban park favoured especially by dog walkers, I skirt the edge of the Lourens River, with its population of self-sustaining trout, and often mentally fish it: the trout down here are as 'wild' as those in the sections of the river I fish further upstream, which are 'closed' to the general public – maybe more so as they face not only the usual predation by birds and animals, but also frequent disturbance by overfed labradors disporting themselves in the water. If one aspect of the wildness of species might be their resistance to capture or their wariness of humans, Harold F. Blaisdell points out that ironically fish in 'unfished wilderness waters' are frequently 'gullible and unsophisticated'.[23] In contrast, heavily fished waters, especially those that enforce catch and release only, are often described as holding 'educated fish'.

* * *

We are having a tapas lunch at Fork Restaurant in Long Street, sitting on the balcony, sipping wine and watching the street life below. Inexplicably, a gymnogene alights on the balustrade before disappearing under the eaves of the next building where we assume it has a nest. 'Wow, look at that fucking great hybrid, feral pigeon!' exclaims one of the diners.

* * *

The poet, environmental scholar and Buddhist thinker Gary Snyder takes the concept of wildness even further in his book *The Practice of the Wild*. 'Wildness is not just the "preservation of the world", it *is* the world', he argues.[24] For Snyder, attentiveness to wildness is a way of being in the world; it is in fact the very 'etiquette of freedom'.[25] Wildness is, however, an elusive concept. It is, in his wonderfully suggestive simile, 'like a gray fox trotting off through the forest, ducking behind bushes, going in and out of sight'.[26]

Snyder points out that 'wild' is usually defined by what it is not: not tame, not cultivated, not inhibited, uncivilised, and so on. But then he turns this argument back on itself, asking what it would mean to define it not as a negative but a positive value (self-propagating, growing sustainably, self-reliant, independent, and so on), as a way of rethinking who we are and how we relate to other species and the environments we share.[27] I find his approach extremely enabling.

In stark contrast to the US Wilderness Act, which insists on human absence as a precondition for wilderness, Snyder contends that a 'properly radical environmentalist position is in no way anti-human'.[28] However, that is not to imply human-centredness. Instead, the 'critical argument now within environmental circles is between those who operate from a human-centered resource management mentality and those whose values reflect an awareness of the integrity of the whole of nature. The latter position, that of Deep Ecology, is politically livelier, more courageous, more convivial, riskier, and more scientific.'[29] This view insists that 'the natural world has value in its own right, that the health of natural systems should be our first concern, and that this best serves the interests of humans as well'.[30]

Snyder rejects the argument that the long human involvement with environmental processes negates the notion of the wild: 'The presence of human beings does not negate wilderness. It's a matter of how much wildness as process is left intact.'[31] In the book *Are Trout South African? Stories of Fish, People and Places*, I referred to the conception that if 'nature' is defined as the absence of human influence, in one sense there can be 'no more nature' in that by changing the weather patterns of the planet, humans have changed everything.[32] Neither Snyder's nor my argument, however, means that concepts such as 'wild' or 'nature' are not useful. They are. But they are of little or no use as absolutes.

Snyder's position is more radical than any of those considered so far, in that he construes 'being human' as at a fundamental level also 'being wild'. Even the very term generally opposed to wildness or naturalness – culture – has, he points out, at its heart a duality: its reference to patterns, structures, habits and products of human behaviour, or, following one part of Raymond Williams's definition, 'a whole way of life', is 'never far from a biological root meaning as in "yoghurt culture" – a nourishing habitat'.[33] He also observes that human beings 'are still a wild species [in that] our breeding has never been controlled for the purpose of any specific yield'.[34] Culture itself is seen to have a 'wild edge', and Snyder quotes Claude Levi Strauss's observation that 'the arts are the wilderness areas of the imagination surviving, like national parks, in the midst of civilized minds. The abandon and delight of love-making is, as often sung, part of the delightful wild in us.'[35]

Our bodies themselves are wild, says Snyder, revealing 'the universal responses of this mammal body': 'The involuntary quick turn of the head at a shout, the vertigo at looking off a precipice, the heart-in-the-throat in a moment of danger, the catch of the breath, the quiet moments relaxing, staring, reflecting.'[36] And he points out that 'the world' itself, 'with the exception of a tiny bit of human intervention, is ultimately a wild place. It is that side of our being which guides our breath and our digestion, and when observed and appreciated is a source of deep intelligence.'[37] In this respect, he refers to Henry David Thoreau's '"awful ferity" shared by virtuous people and lovers'.[38]

For Snyder, wildness is the norm, not the exception: 'It has always been part of basic human experience to live in a culture of wilderness. There has been no wilderness without some kind of human presence for several hundred thousand years. Nature is not a place to visit, it is *home* – and within that home territory there are more and less familiar places.'[39] He does acknowledge, though, that many of us have lost touch with the wild, including that within ourselves. Wildness is also ubiquitous, and not restricted to so-called 'wilderness areas', which make up 2 per cent of the land mass of the US, for example:

Shifting scales, it is everywhere: ineradicable populations of fungi, moss, mold, yeasts and such that surround and inhabit us. Deer mice on the back porch, deer bounding across the freeway, pigeons in the park, spiders in the corners . . . Exquisite complex beings in their energy webs inhabiting the fertile corners of the urban world in accord with the rules of wild systems, the visible hardy stalks and stems of vacant lots and railroads, the persistent raccoon squads, bacteria in the loam and in our yogurt.[40]

Wilderness areas are simply those places in which 'the wild potential is fully expressed, a diversity of living and nonliving beings flourishing according to their own sorts of order'.[41] But, as the quotation by Snyder above suggests, spiders, lizards, owls, mice, goshawks, mosquitoes and an almost endless list of other species manage to exist 'according to their own sorts of order' even in suburbia, so cities are also environments of potential wildness, something I explore in more detail in a later chapter.

In conversation with Jim Harrison in the film *The Etiquette of Freedom*, Snyder takes this idea of wildness as quality, as characteristic that can exist in gradations, further in introducing the term 'working landscapes':

GS: [O]ne of the terms I find myself using more now is 'working landscapes', to be distinguished from the idea of totally pristine wilderness landscapes. And that's what we have here along the California coast – a lot of working landscapes. The wild works on all scales.

JH: Yeah. Some of the wildest places in England are the old Roman cart trails that have been eroded. They're called 'hollow ways', twenty feet deep; they've become these preposterously dense thickets.

GS: And that's part of it – the wild can be a wood lot. Even the vacant lot in the city can be wild.[42]

Snyder argues that 'civilization is ego gone to seed and institutionalized in the form of the State, both Eastern and Western', and that 'there is an almost self-congratulatory *ignorance* of the natural world that is pervasive in Euro-American business, political, and religious circles'.[43] (I would add that that ignorance is widely evident across the African continent too, nowadays, especially with the increasing shift to urban or peri-urban dwelling as the norm.) Accordingly, one of Snyder's imperatives over some years has been running workshops that teach 'the disciplines, knowledge and skills . . . necessary to appreciate the ferocious orderliness of the wild'.[44] Hence his inclusion of the word 'practice' in his book title. We need to learn and relearn through practice the etiquette required by a system – the wild – which is 'actually impartially, relentlessly and beautifully formal and free'.[45] This involves an 'ethical life . . . that is mindful, mannerly and has style'; which eschews 'stinginess of thought, which includes meanness in all its forms';[46] and which requires 'not only generosity but a good-humoured toughness that cheerfully tolerates discomfort, an appreciation of everyone's fragility, and a certain modesty'.[47] It is a life 'vowed to simplicity, appropriate boldness, good humour, gratitude, unstinting work and play, and lots of walking [that] brings us close to the actually existing world and its wholeness'[48], and which seeks to know the environment intimately. Such a life is not shy of hard work, and involves knowing and enjoying 'the skills of our hands and our well-made tools'.[49]

The etiquette of wildness is as much about human pleasure and joy – an emotional, psychological and spiritual process – as it is about biological necessity:

> The wilderness pilgrim's step-by-step breath-by-breath walk up a trail, into those snowfields, carrying all on the back, is so ancient a set of gestures as to bring a profound sense of body-mind joy. Not just backpackers, of course. The same happens to those who sail the oceans, kayak fjords or rivers, tend a garden, peel garlic . . . The point is to make intimate contact with the real world, the real self.[50]

And we should return from such experiences to 'tell a good story'.[51] Snyder stresses, however, that one does not abandon the mess of the everyday for the wildness of the 'out there', but that one reads the everyday as part of the continuum with the wild:

> One should not dwell in the specialness of the extraordinary experience nor hope to leave the political quag behind to enter a perpetual state of heightened insight. The best purposes of such studies and hikes is to be able to come back to the lowlands and see all about us, agricultural, suburban, urban, as part of the same territory – never totally ruined, never completely unnatural.[52]

Snyder points to the fallacy, frequently expounded by scholars, environmental activists and members of so-called 'indigenous societies', that hunter-gatherer communities lived very lightly on the land, and did not affect its processes at all. I pointed to refutations of this argument earlier in relation to Australia. Snyder discusses Britain and Guatemala:

> We also know that early economies often were more manipulative of the environment than is commonly realized. The people of Mesolithic Britain selectively cleared or burned the valley of the Thames as a way to encourage the growth of hazel. An almost invisible system of nut and fruit growing was once practised in the jungles of Guatemala.[53]

Further examples could be drawn from Canada, in which Native Canadians have for centuries collected fertilised salmon ova, transported them in boxes lined with wet moss, and then seeded them into other rivers, and have also set fires to ensure grassland stayed open for hunting and to contain the encroachment of forests. In contrast to human efforts that have consistently worked against wild processes (clearing land, channelling rivers, poisoning or pulling up weeds, ploughing the soil), Snyder asks whether there is not the possibility of agriculture 'going *with* rather than against nature's tendency', of '[d]oing horticulture, agriculture, or forestry with the grain rather than against it', acknowledging that 'the source of fertility ultimately is the "wild"', and that 'good soil is good because of the wildness in it'.[54] An example of this approach might be 'no till farming', in which the structure of the soil is not compromised by ploughing.

Can the wild be cruel? In one sense to ask that is either to anthropomorphise animal behaviour or potentially to make a category error in asking a 'human' question of 'non-human' behaviour. A theologian colleague of mine, Ernst Conradie, asked the question in an even more contentious way: Can animals sin?[55] It is a 'cruel' question to ask of anyone, as you will find yourself pondering it endlessly, and as Conradie himself observed, you are in trouble whichever way you answer it. I discuss these issues more extensively in the chapter on 'Wild Ethics'. Snyder does not shy away from the violence and destruction inherent in wildness:

> Life in the wild is not just eating berries in the sunlight. I like to imagine a 'depth ecology' that would go to the dark side of nature – the ball of crunched bones in a scat, the feathers in the snow, tales of insatiable appetite. Wild systems are in one elevated sense above criticism, but they can also be seen as irrational, moldy, cruel, parasitic. Jim Dodge once told me how he had watched – with fascinated horror – Orcas methodically batter a Gray Whale to death in the Chukchi Sea.[56]

There are many stories of predators killing or maiming far more prey animals than they could possibly eat (a practice known as 'surplus killing' or 'henhouse syndrome'), including biting through the Achilles tendons of sheep and leaving the animals to die of starvation, thirst or blood loss. Scientists have argued that this may be intended to provide a food reserve for times when prey is scarce or hunting conditions are poor. Monbiot refers to specific instances of wolves at times killing more than they can eat in one meal, and then returning to the carcasses over periods of some months, almost like visiting a larder.[57] There are, however, instances where much of the dead prey is simply left to rot. One fox in Australia killed eleven wallabies and 74 penguins over a period of days, and ate almost none. A leopard in South Africa killed 51 sheep and lambs in one incident. Two caracal, also in South Africa, killed 22 sheep in one night, but ate only the buttock of one of them. And up to nineteen spotted hyenas are recorded to have killed 82 Thomson's gazelle and badly injured a further 27, though only 16 per cent of the animals killed were actually eaten.[58] The practice is complex and potentially disturbing in its implications.

Intriguingly, unlike many who write on these subjects, Snyder does not propose vegetarianism as the only ethical option for human-animal relations.[59] For Snyder, killing in order to eat is a reality with which we must properly come to terms, and he praises subsistence economies as 'sacramental' in that they have 'faced up to one of the critical problems of life and death: the taking of life for food'.[60] In contrast, in consumer economies '[o]ur distance from the source of our food enables us to be superficially more comfortable, and distinctly more ignorant'.[61] His relative comfort with the taking of life for food – the acknowledgement that 'each of us at the table will eventually be part of the meal'[62] – appears also to flow from his belief in the spiritual interconnectedness of beings, in which animals and plants do not mind if we eat them, provided that we do so with respect and gratitude. I do not follow Snyder (or Monbiot, who does something similar) down this spiritual path.

In describing his engagement with wildness, Snyder uses phrases like being 'born again', experiencing a 'worldwide purification of mind', being 'advised by a tree',[63] and he refers to talking to rocks, trees and even on one occasion a prayer hut that had been transported from India. This emerges, I assume, from his own immersion in Zen Buddhism, but is to my mind extraneous to his argument about wildness, binding the argument to one belief system whereas it is potentially of value to many. Again, in terms of his own Buddhist beliefs, Snyder claims at various stages that there is a necessary causal connection between monotheism and biological monoculture.[64] I do not agree with him. To use the example of Christianity, a reading of the account of creation in Genesis, and of God's response to Job at the end of that book, indicates a revelling in the multiplicity and difference of life forms, very far removed from monoculture. Rather, it might be more accurate to say that capitalism promotes biological monoculture, to the extent even of modifying genetic structure.

* * *

There was a gentle hubbub on the beach: an ice-cream van, a handful of cars, some children wading and splashing in the narrow runnels trapped by the sandbars when the tide had pulled the plug. Beyond the cars I saw a wonderful sight. An ancient woman wearing iridescent ski goggles and a blanket over her knees was riding her electric wheelchair at full tilt. Sand spurted from the wheels. She skidded around in tight circles, jolted forward and fishtailed through the ruts left by the cars. Someone's heart was still beating. — George Monbiot, Feral

* * *

The 2015 documentary film *Salmon: Running the Gauntlet* by National Geographic/BBC examines attempts to restore the major salmon runs in North America, which have all but come to a halt in many rivers. The film points out that the returning salmon constitute an enormous food resource for bears, bald eagles, ospreys, cormorants, humans, and a whole range of others. The salmon runs also transfer a rich range of nutrients from the Pacific Ocean into inland areas that are otherwise relatively infertile. These nutrients are then spread into land surrounding the rivers by animals and birds that eat the fish and then defecate in the forests and surrounding areas. Even a ponderosa pine growing some distance from the river, for example, will absorb nutrients from the salmon. One study indicated that 'between 15 and 18 per cent of the nitrogen in the leaves of spruce trees within 500 metres of a salmon stream comes from the sea: it was brought upriver in the bodies of the salmon'.[65]

The salmon runs have been badly affected by overfishing, habitat degradation and the damming (damning?) of rivers. The film documents human efforts to restore the salmon runs by catching the few salmon that do return to their native breeding sites, stripping them of their eggs and milt, and then hatching the fertilised eggs in hatcheries. Hundreds of millions of salmon fry are then stocked into the rivers, but only about 1 per cent return to spawn.

The reasons for the poor return are various. Many of the salmon runs have suffered because large dams have been built on the rivers, effectively barring the fish from access to areas upstream. Some dams have installed salmon ladders to assist the fish, but the congregation of fish below these ladders makes them extremely vulnerable to predators, especially seals in the lower (estuarine) reaches. The US Wildlife Department actually employs people to fire rubber bullets and other 'non-lethal' projectiles to keep the seals at bay in such instances. These dams also cause other problems, though. Salmon fry return to the sea not by swimming back down the river, as one might assume, but by facing upstream and allowing the current to sweep them downstream. Swimming across vast bodies of still water (the dams or the rivers upstream or downstream from them) exhausts the small fish, which increases the mortality rate. In many cases, they also have to survive a precipitous drop of up to 100 metres or more when they reach the dam wall, which kills many or disorients them, leaving them easy prey for predators. Attempts have been made to avoid this problem by capturing the salmon fry before they reach these obstacles and transporting them downstream via truck or purpose-built barge, and then releasing them lower down. However, predatory birds, such as terns and cormorants, have quickly wised up to this fast-food delivery system, and established large breeding colonies at sites where the fish are dumped. In some rivers, the ready supply of small salmon has led to an increase in the populations of predatory fish such as the pikeminnow, to the extent that there is now a state bounty for each of these fish caught and killed. In the documentary, one angler admitted that the financial reward for catching these fish was such that he shuts his contracting business for three months of the year just to earn money killing pikeminnows. These are all problems that had been identified as early as 1959 by the renowned judge, fly-fishing author and environmentalist Roderick Haig-Brown, when he warned in his book, *Fisherman's Summer*, that salmon runs and hydroelectric dams could never coexist, no matter the technology employed to transport the salmon over or around walls.[66] He was chillingly correct.

But perhaps the most serious problem with the captive breeding project is the reduction of genetic diversity. In 'natural' or 'wild' salmon runs, the gene pool is massive, and only the genetically strongest fish would survive to return to spawn again. The successful hatching rate of ova in the wild is very low. In hatcheries, the eggs and milt are drawn from a comparatively small number of fish, and the high rate of production results from the extremely high rate of successful hatching. So a very large number of genetically similar fish (which in the bulk of cases would have not hatched in the 'wild') is introduced into the river system. The costs have been enormous, and the results spectacularly unsuccessful.

More successful have been recent developments, including the destruction of dams, and also efforts to restore rivers that have been canalised and so offer few resting areas for migrating fish, by introducing structure in the form of felled trees and rocks protruding into the current. In cases like this, the salmon have 'naturally' started to return. Even so apparently 'simple' a solution as a legal judgment that required a major dam to release water while the salmon fry were moving downstream, so as to ensure that they could progress down with the current, facing upstream, increased the returning run substantially.

The salmon hatchery breeding project was hailed at the time in messianic terms, as promising an almost endless supply of salmon in a context in which natural stocks had been all but annihilated by overfishing. It seemed a triumph of human scientific ingenuity. Besides providing hatchery-reared salmon for the table, however, its efforts have yielded very little, and the biological pollution (disease, excess feed and faeces) resulting from many salmon farms has been extensively documented. Where we have allowed wildness to return, where we have 'rewilded' the rivers, the results have been remarkably better.

* * *

We modern humans, increasingly competent about making our way through the natural world, have been decreasingly confident about its values, its meanings. The correlation is not accidental. It is hard to discover meaning in a world where value appears only at the human touch, hard to locate meaning when we are engulfed in sheer instrumentality, whether of artifacts or natural resources. One needs a significant place to dwell. — Holmes Rolston III, Environmental Ethics

* * *

Monbiot's book, *Feral: Rewilding the Land, Sea and Human Life*, is an extraordinary account of 'rewilding'[67] as a biological, ethical and imaginative imperative. Like Snyder, Monbiot rejects the idea of a romanticising call for a return to a hunter-gatherer lifestyle 'in tune with nature', pointing out that he does not 'romanticise evolutionary time':

> I have already lived beyond the lifespan of most hunter-gatherers. Without farming, sanitation, vaccination, antibiotics, surgery and optometry I would be dead by now. The outcome of mortal combat between me, myopically stumbling around with a stone-tipped spear, and an enraged giant aurochs is not hard to predict.
> The study of past ecosystems shows us that whenever people broke into new lands, however rudimentary their technology and small their numbers, they soon destroyed much of the wildlife – especially the larger animals – that lived there. There was no state of grace, no golden age in which people lived in harmony with nature.[68]

Arguing that there is abundant evidence for the environmental impact of hunter-gatherers in North America, Monbiot cites Ran Prieur on the fact that 'passenger pigeons were not even common in the 1400s', and that such people 'specifically targeted pregnant deer and wild turkeys before they laid their eggs, to eliminate competition for maize and tree nuts'; that they burned forests for their purposes; and they impacted the populations of species that they ate, including salmon and crayfish, thereby also affecting the other species that predated upon them.[69]

I referred earlier to the phenomenon of 'surplus killing' or 'henhouse syndrome' among animal predators. The accounts of surplus killing among colonial settlers are extremely well known, if no less disturbing. But Monbiot makes the telling point that '[h]ad the Mesolithic people of the Americas eaten everything they killed, they would scarcely have trimmed the herds of game, so small were their numbers'. He points out that one ground sloth, which weighed the equivalent of twenty elephants, could have sustained a group of hunter-gatherers for months. 'The speed with which the megafauna of the Americas collapsed might suggest that they slaughtered everything they encountered.' He suggests: 'Perhaps the care with which some indigenous people of the Americas engage with the natural world came later.'[70]

Like that of Snyder, Monbiot's engagement with the environment is as much existential as it is biological (in fact I assume both thinkers would insist that the two cannot be separated). Monbiot confesses to realising that he was in a state of 'ecological boredom'. It was not that he was lacking 'authenticity', he argues, as he does not find this a useful concept; rather he 'wanted only to satisfy [his] craving for a richer, rawer life than [he] had recently lived'.[71] If Snyder sometimes verges on the missionary in his arguments about 'wildness', Monbiot is more pragmatic in acknowledging: '[S]omehow I had to reconcile this urge with the life I could not abandon: bringing up my child, paying my mortgage, respecting the rights and needs of other people, restraining myself from damaging the natural world.' He goes on: 'It was only when I stumbled across an unfamiliar word that I began to understand what I was looking for': the word was 'rewilding'.[72] Rewilding is not a retreat from nature, but a 're-involvement [. . .] an enhanced opportunity for people to engage with and delight in the natural world'.[73]

Rewilding is not about restoring ecosystems (though it can involve the reintroduction of species), but rather about 'permit[ting] ecological processes to resume'.[74] Monbiot points out that many well-intended conservation efforts attempt to restore, or freeze in time, ecosystems that were already the product of environmental degradation, and that these involve the deliberate and continued barring of access or egress to particular species: 'It is as if the conservationists in the Amazon had decided to protect the cattle ranches rather than the rainforest.'[75] Rewilding is an approach that recognises that 'nature consists not just of a collection of species but also of their ever-shifting relationships with each other and with the physical environment'. He argues against environmental models that seek to keep an ecosystem in a state of arrested development, preserving it 'as if it were a jar of pickles' and protecting 'something which bears very little relationship to the natural world'.[76]

This involves a far longer view, and one that understands the fundamental point, made also by the fly-fishing author Paul Schullery,[77] which I take up in a later chapter, that environmental processes are ongoing, and that the natural world 'is never a finished product'. Monbiot makes the rather surprising observation that scientific evidence suggests that many tree and shrub species in Europe developed the characteristics they have to defend themselves against the 'straight tusked elephants' that lived in Europe until about 40 000 years ago.[78] He also points out that there were beavers in Britain until the mid-eighteenth century, and that hippo remains have been found under Trafalgar Square (probably dating back 100 000 years).[79] The presence of bears in the past might seem unsurprising; lions in Britain and Europe perhaps more so.

What seems to me Monbiot's most definitive statement on rewilding is worth quoting in full:

I have no desire to try to re-create the landscapes or ecosystems that existed in the past, to reconstruct – as if that were possible – primordial wilderness. Rewilding, to me, is about resisting the urge to control nature and allowing it to find its own way. It involves reintroducing absent plants and animals (and in a few cases culling exotic species which cannot be contained by native wildlife), pulling down the fences, blocking the drainage ditches, but otherwise stepping back. At sea, it means excluding commercial fishing and other forms of exploitation. The ecosystems that result are best described not as wilderness, but as self-willed: governed not by human management but by their own processes. Rewilding has no end points, no view about what a 'right' ecosystem or 'right' assemblage of species looks like. It does not strive to produce a heath, a meadow, a rainforest, a kelp garden or a coral reef. It lets nature decide.[80]

Monbiot quotes Alan Watson Featherstone, who heads up the organisation Trees for Life, as pointing out that rewilding is 'crucially different from the ethos of human domination . . . [It] is about humility, about stepping back'.[81] As I pointed out in relation to debates about ecological restoration in Australia, climatic and geomorphological conditions are very different now than they were at whatever historical period is chosen as the model for recreation. In terms of rewilding, this also means that the environments that result from this approach may be different from those of the past.[82]

Monbiot stresses that rewilding is not an approach that should be applied in blanket fashion. Each context must be carefully evaluated. He cites an example from the Russian tundra in which rewilding would likely lead to the displacement of moss by grasses, releasing powerful greenhouse gasses that would be damaging.[83] In contrast to the view of 'primitivists', rewilding does not perceive a conflict between the 'civilised' and the 'wild'. Monbiot argues that we can 'enjoy the benefits of advanced technology while also enjoying, if we choose, a life richer in adventure and surprise. Rewilding is not about abandoning civilization but enhancing it.'[84] In fact, he points out that we cannot abandon 'a sophisticated economy, supported by high crop yields' because the consequences for humans would be disastrous. He calculates that before the development of farming, Britain probably supported 5 000 people in total, 'one per 54 square kilometres'. Monbiot notes: 'The fantasy entertained by some of the primitivists . . . of returning to a hunter-gatherer economy, would first require the elimination of almost all human beings.'[85] Accordingly, his suggestions for what kinds of land should be rewilded are quite modest, but they nevertheless constitute an important ecological and existential resource.

In a way similar to Snyder, Monbiot is concerned with human identity, well-being and adjustment as part of a re-engagement with biological processes and landscapes. Rewilding is an ecological and human-emotional-psychological-physiological imperative, and Monbiot says that his book 'also examines the lives we may no longer lead and the constraints – many of them necessary – that prevent us from exercising some of our neglected faculties. It explains how I have sought, within these constraints, to rewild my own life [. . .].'[86]

Whereas environmental discourse has tended to be negative – we are told what we should not do or stop doing – rewilding has the potential to suggest what we should do, to propose a 'positive environmentalism'.[87] Monbiot points to the tragic irony that '[m]ost of the rewilding that has taken place on earth so far has happened as a result of humanitarian disasters'.[88] A case in point would be the substantial regeneration of fish stocks along the east coast of Africa in recent years because attacks by Somali pirates have kept fishing boats out of the area. A more extreme (and sinister) example might be the abandoned site of the Chernobyl nuclear disaster, from which humans have had to withdraw completely. But Monbiot is adamant that planned rewilding must never take place at the expense of humans or by coercive means. He is critical of projects in East Africa and Botswana that have created 'a paradise for the rich on the land of the poor', commenting that '[i]f a rewilding scheme requires forced dispossession, it should not go ahead'.[89] It is a point he emphasises several times in the book: 'It should happen only with the consent and enthusiasm of those who work on the land. It must never be used as an instrument of expropriation or dispossession.'[90]

He also stresses that if rewilding threatens livelihoods, as in the case of a reintroduced wolf repeatedly killing sheep, for example, the animal could be legitimately shot: 'I think we should be able to love wildlife without being unreasonably sentimental,' he argues.[91] In drawing up a list of animals that could be reintroduced to Britain, he specifically takes into account the potential of species to threaten humans or their livelihoods, and recommends against their being part of such a project.[92] He also argues that rewilding should not be done on viable agricultural land, and that it is a resource for those with an affinity for such experiences, not something compulsory or enforced. I am in full agreement. Personally, I would opt on almost every occasion for an afternoon on a beach, by a river or in a garden over one spent in a shopping mall, but I have several friends and colleagues who would choose the opposite, as is their right.

Feral focuses most of its attention on the United Kingdom, and Monbiot expresses despair that so many of its conservation bodies celebrate moorlands and heath as areas of wildness and wilderness. These landscapes, he points out, are the result of systematic clearing of the woodland, and even rainforest, which covered almost all of Britain by farmers, and especially by sheep (he has a particular ferocity towards sheep). This process began with the farmers of the Neolithic period (between 4 000 and 6 000 years ago), and continues to this day, including in the rather misguided efforts of conservation bodies to preserve moors and grasslands by excluding or removing trees and bushes. He points out that

> the open landscapes of upland Britain, the heaths and moors and blanket bogs, the rough grassland and bare rock which many see as the natural state of the hills, which feature in a thousand romantic films and a thousand advertisements for clothes and cars and mineral water, are the result of human activity, mostly the grazing of sheep and cattle . . . [G]razing and cultivation have depleted the soil . . . [W]hen grazing pressure eases, trees can return.[93]

In this respect, Monbiot introduces Daniel Pauly's concept of the 'shifting baseline syndrome', according to which each generation assumes that the environmental conditions that they encountered as children were 'normal', and thus form the basis of what we should attempt to restore. So, with each generation, people establish their own ecological baseline, often unaware, Monbiot points out, that 'what they considered normal when they were children was in fact a state of extreme depletion'.[94] In this regard, I read with interest of efforts to conserve about fifteen hectares of an endangered vegetation type known as Lourensford alluvium fynbos (LAF) on Vergelegen wine farm near my home. This stretch is 'considered the most conservation-worthy section of this vegetation type in the world'.[95] But the joy of the discovery is tempered by the acknowledgement, contra-shifting baseline syndrome, that LAF originally covered about 6 000 hectares in the region, and that only 9 per cent remains.

I mentioned that Monbiot has a particular ferocity for sheep, which he describes as 'the white plague':[96]

> Because they were never part of our native ecosystem, the vegetation of this country has evolved no defences against sheep. In the upland they rapidly deplete nutritious and palatable plants, leaving behind a remarkably impoverished flora: little besides moss, moorgrass and tormentil in many places. The sheep has caused more extensive environmental damage in this country than all the building that has ever taken place here.[97]

In case we have not grasped the point, he reiterates later in *Feral*: 'Sheep farming in this country is a slow-burning ecological disaster, which has done more damage to the living systems of this country than either climate change or industrial pollution. Yet scarcely anyone seems to have noticed.'[98] He argues that the unquestioned primacy of the interest of famers and landowners in public debate and policy (he calls it 'agricultural hegemony') means that sheep have 'full diplomatic immunity'.[99] His comments raise the interesting question about the relative absence of arguments about the impacts of farm animals on the environment in South Africa, with the exception of erosion caused by overgrazing or by goats, and the generalised comments that the greenhouse effect globally is to a large extent the result of the methane produced by farting cows.

Unlike the ecological baselines established by successive generations, rewilding does not have fixed objectives, as in a clear idea of what environment it seeks to create; instead 'the process is the outcome'. Monbiot claims: 'The main aim of rewilding is to restore to the greatest extent possible ecology's dynamic interactions. In other words, the scientific principle behind rewilding is restoring what ecologists call trophic [related to food and feeding] diversity.'[100] It means 'enhancing the number of opportunities for animals, plants and other creatures to feed on each other [. . .] expanding the web both vertically and horizontally, increasing the number of trophic levels (top predators, middle predators, plant eaters, plants, carrion and detritus feeders) and creating opportunities for the number and complexity of relationships at every level to rise'.[101]

In contrast to the bottom-up approach previously adopted by scientists, possibly resulting from studying already depleted ecosystems and according to which numbers of predators are controlled by the numbers of their prey or the abundance or otherwise of the vegetation on which the prey feeds, those studying ecosystems now think in terms of 'trophic cascades', which occur 'when the animals at the top of the food chain – the top predators – change the numbers not just of their prey, but also of species with which they have no direct connection. Their impacts cascade down the food chain, in some cases radically changing the ecosystem, the landscape and even the chemical composition of the soil.'[102] The impact of salmon runs, even on trees some distance from the river itself, is an example of trophic cascade. But probably the best-known example is the reintroduction of wolves to Yellowstone Park, which transformed for the better 'almost every aspect of the ecosystem', even altering 'the physical geography of the site, changing the flow and shape of the rivers and erosion rates of the land'.[103] To explain briefly, the deer now had to stay closer to cover to avoid predation, so they did not graze right up to the river banks, which were thus protected from erosion, leading to cleaner water, better flow, and more bank structure for fish to use for cover.

As a significant, though parenthetical, observation, the environmental lawyer Anel du Plessis, who was a research fellow at the Stellenbosch Institute for Advanced Study at the same time I was there working on the early stages of this project, pointed out to me that the bulk of the disused mines in South Africa cannot officially be closed because they cannot be returned to a state resembling what the site was like previously, as a result both of the potential cost involved and the extent of damage caused during the mining operations: they cannot be sufficiently rewilded.

In a discussion that is cogent to the South African context, or to the ecological history of just about anywhere characterised by the establishment of species 'from elsewhere', Monbiot ponders the question of rewilding in the US, using species from elsewhere that are similar to those that have disappeared: 'Is a healthy and desirable ecosystem necessarily composed of native species?'[104] He points out that many plants that are very familiar to people living in the UK, up to 157 by some estimates, which were previously regarded as native, 'now appear to be what botanists call archaeophytes: exotic species which arrived before 1500'.[105] Does it matter? As I argued in *Are Trout South African?*, to my mind only if they cause harm and do not add value of some kind. Monbiot agrees: 'Restoring a functioning ecosystem does not equate to purging all non-native species. It requires only that we control or suppress those species which deprive many others of a foothold there.'[106]

Feral includes many accounts of Monbiot's own personal experiences of activities, such as fishing, kayaking, walking, and so on, and towards the end he recounts the transformation that has occurred within himself:

It was then that I realized that a rewilding, for me, had already begun. By seeking out the pockets of land and water that might inspire and guide an attempt to revive the natural world, I had revived my own life. Long before my dreams of restoration had been realized, the untamed spirit I had sought to invoke had already returned. By equipping myself with knowledge of the past while imagining a rawer and richer future, I had banished my ecological boredom. The world had come alive with meaning, alive with possibility.[107]

It is a realisation with which many who seek out wild places and experiences – those who fish, hike, photograph landscapes, collect mushrooms, climb mountains, canoe, sail, surf – can identify. But, like Snyder, Monbiot sometimes crosses into narrative terrain in which 'the wild' is either part of primordial human spirituality or else primordial genetic memory. Here are two examples. In the first, he narrates trying to spear a flounder:

As I stalked up the channel, my spear poised above the water, I felt as flexed and focused as a heron. Every cell seemed stretched, tuned like a string to the world through which I moved, straining for a note among the shifting harmonics of wind and water. My concentration intensified until I became hyper-aware, sensing each grain beneath my bare toes, every ripple round my waist, every movement, however infinitesimal, among the benthos. Suddenly I was gone.

It is hard to explain what happened. Perhaps it was the mesmeric repetition of the ripples in the sand, perhaps an escalating pitch of attention that thrust me through the barrier of the present, but I was at that moment transported by the thought – the knowledge – that I had done this before.

Except for the two forays I have mentioned already [earlier in the book], I had not. I do not believe in reincarnation, or in the persistence of the soul after the death of the body. Yet I felt I was walking through something I had done a thousand times, that I knew this work as surely as I knew my way home.[108]

In the second example, while foraging for herbs and fungi, he finds a small deer that has just died, and decides to sling it over his shoulders and take it home to eat:

> The deer wrapped around my neck and back as if it had been tailored for me; the weight seemed to settle perfectly across my joints. The effect was remarkable. As soon as I felt its warmth on my back, I wanted to roar. My skin flushed, my lungs filled with air. This, my body told me, was why I was here. This was what I was for. Civilization slid off as easily as a bathrobe.[109]

He does qualify this account with the self-deflating footnote: 'Picking up an animal that has died of natural causes and taking it home is a foolish thing to do: when I phoned a veterinary surgeon I know to ask if I could eat the deer, he told me to bury it.'[110] But then he adds: '[T]his sensation was new. I could not assimilate it because – until I picked up the deer – I had been unaware of its existence. It was overwhelming, raw, feral. I did not have a place to put it; but I knew that it belonged to me as much as the tendons I use to curl my fingers.' He attributes this to 'genetic memory': 'These genetic memories – these unconsidered urges – are printed into our chromosomes, an irreducible component of our identity.'[111] I am not sure that I would want to go down that line of argument, nor do I think it is particularly helpful.

*   *   *

'Wilderness Area' proclaims a sign beside an unremarkable field outside the village of Darling. For a second I think someone has a finely absurd sense of humour. Then I realise I am looking at a wildflower reserve, in which the composition of the veld is sufficiently intact to allow the springtime germination of the famous West Coast wild flowers. I look again, with new eyes, at several hectares of grasses procreating in quiet abandon.

*   *   *

In the chapters that follow, I explore these broader ideas about wildness and rewilding in relation to key aspects of the way in which we live in the world, building an argument for the possibilities of other ways of living, both individually and as a species. I discuss wildness and conservation; wildness and local language; wild seams and margins; wild cities; the wild and the farmed; wild fish; and wild ethics. I do not try to offer a definitive account or exhaustive set of arguments in the chapters. Wildness simply exceeds our attempts to describe it, which is part of its allure. The chapters are rather intended to be suggestive, as spurs to further thought and engagement on the part of readers, and cumulatively to sketch the possibilities of 'wilder lives'. Rewilding – including of one's own life – is a lifelong commitment.

3

Wildness and Conservation

Does the hawk take flight by your wisdom
    and spread his wings toward the south?
Does the eagle soar at your command
    and build his nest on high?

— Job 39: 26–27 NIV

Tourists visiting South Africa will frequently talk of their visit to a
game park or lodge as 'going on safari'. The *Oxford English Dictionary*
defines the term 'safari' as 'an expedition to observe or hunt animals in
their natural habitat, especially in East Africa', and notes its origin in
the late nineteenth century from the Arabic, via Kiswahili, meaning
'to travel'. It is a term of colonial origin, connoting a brave journey
into untamed wilderness, and an encounter with wild animals, and is
much favoured by tourist franchises catering to the overseas market.
Anyone walking through a South African airport will notice that
tourists 'on safari' will frequently wear for the occasion multi-pocketed
'safari jackets', rather like those used by photographers, hunters or fly
fishers. Alongside ensuring that every creature comfort is catered for,
game parks and lodges are at pains to maintain the illusion of wildness,
of 'on safariness'.

For South Africans, the term 'safari' might suggest, instead, a
packet of dried fruit, a box of firelighters, an off-road Nissan vehicle,[1]
or if they are old enough, the fashion abomination known as the 'safari

suit', especially the version involving short pants worn with long socks. Rather than 'going on safari', they would probably say that they were going to a game reserve or game lodge, or simply use the name of the reserve as shorthand – Kruger, Umfolozi, Hluhluwe, and so on.

Nevertheless, the attraction of seeing 'wild animals' is central to the experiences of both groups, however they may describe what they are doing. This is, of course, a positive thing for the animals and the ecosystems that they inhabit, in that human valuing of and investment in healthy and functioning biological systems is essential for their futures. And the experience of hearing a lion roar into the night close by is a significant, and humbling, one for the mammalian soul.

In this chapter, I look at some of the dynamics at play in 'conserving wildness', whether through rewilding agricultural land or the establishment and management of wildlife reserves, both for animal and plant species and for human communities. These dynamics are often framed as seeking to balance both 'brown' (economic and social developmental) and 'green' (environmental) agendas.

* * *

A family we know has a springbok skin as a mat in the entrance hall of their home. A friend of their son's looked at it and asked, utterly aghast, 'Oh no, what happened to your dog?'

* * *

In an article entitled 'Return of the Wild' in the South African edition of the magazine *Country Life*, journalist Andrea Abbott writes of 'rewilding' projects taking place in game parks in South Africa. She quotes biologist and wildlife manager, Ignacio Jiménez Pérez, who was visiting South Africa to learn about 'rewilding' projects on behalf of Tompkins Conservation, as saying: 'In rewilding, South Africa is the light of the world.'[2] The article focuses predominantly on the private reserve, Phinda, which initially comprised 13 000 hectares of 'degraded farmland', but was rewilded using an aerial photograph from 1930 as a template. The transformation involved removing plantations and invasive plants, and then the gradual reintroduction of species. 'Rewilding is transformational,' says the conservation director, 'and the reintroduction of game, specifically buffalo, lion, wild dog and elephant, is restoring more original trophic cascades that start at the top of the food chain and tumble to the bottom.'[3]

'Rewilding' is an extremely important conservation principle, but the 'rewilding' of agricultural land, in which commercial farming companies or private landowners either partially or completely convert land that had previously been utilised for crops or livestock to the farming of game, is a process that is far more complex and mired in controversy than the *Country Life* article may suggest. It is frequently presented as an ecologically responsible conservation strategy, in which land is restored to its 'natural' state and utilised in ways that require far less water, and few, if any, fertilisers or pesticides. There are many examples of farmlands that were marginal for agriculture in the first place, and had limped along financially with severe ecological impacts through water extraction from rivers and erosion from overgrazing, which are now far more secure financially and ecologically having been turned to game. Land value and financial yield for the farm owner typically increase with the conversion to game, and land usage is undoubtedly 'healthier', though the ethics of some of the resultant game farming, conservation or hunting operations may be questionable.

As animals typically move with the seasons, ensuring year-round hunting generally requires the erection of substantial fences, often electrified, to contain the animals' movements. Game animals are of greater monetary value than cows or sheep, so the fences also protect financial interests. The size and location of the farm, and also the decision of what species to reintroduce, including the presence or absence of predators, influence the extent to which the area operates largely in 'wild' terms, or under a fairly tight management regime, though even in the latter case the behaviour of the animals themselves may be decidedly 'wild'.

The impact on local communities is, however, the most controversial and contested aspect. We may recall George Monbiot's insistence that 'rewilding' never be undertaken in a way that would displace people or threaten their livelihoods. That has unfortunately not been the case in South Africa or several other countries. In an article entitled 'Private Game Farming and its Social Consequences in Post-Apartheid South Africa', Marja Spierenburg and Shirley Brooks discuss what they refer to as 'the commodified production of nature and "wilderness"', which is part of a broader 'deagrarianisation' within the African continent and elsewhere.[4] They note that while the shift to what is often called 'wildlife ranching' has been occurring for more than half a century already, the process speeded up considerably post-1994. By 2006, 'wildlife was being produced on nearly 10 000 commercial farms (on about half of which, wildlife is combined with continued crop and/ or livestock activities)'. They note: 'The exact land surface involved in private wildlife production and tourism is difficult to ascertain, but estimates in 2006 varied from 13% to 17% nationally.'[5] When driving through the Eastern Cape or north-eastern Cape, the extent of farm conversion to game is clearly visible. What would previously have been a continuous line of barbed wire fencing, just over a metre high, running for hundreds of kilometres, is now a continuous line of game fencing of 2 metres or more. Twenty to thirty years ago someone driving from Port Elizabeth to Grahamstown would have passed one mixed crop or cattle farm after another. Now there are game lodges, interspersed with game farms, lining the main road on both sides for the entire journey.

Spierenburg and Brooks are concerned with the human costs of the conversion to wildlife ranching. They point out that local people who live on farms very often have long family histories of dwelling on the land, and the farms are for them 'home'. To recognise these histories and attachments, Spierenburg and Brooks suggest a shift in terminology from 'farm workers' to 'farm dwellers'.[6] However, the conversion to game farming has frequently seen people moving or being moved off the land, because game farming is less 'labour intensive', because of the dangers of living with large wild animals, or because the association of wilderness and wildness with 'pristine' areas 'devoid of human presence and influence' means that people are perceived as undesirable, and the land becomes increasingly 'inaccessible to local residents'.[7]

Defenders of the game farming industry tend to draw on arguments framed by the Community-Based Natural Resource Management approach, which claim that these less resource-intensive farming practices can have important benefits to local communities, especially in terms of tourism and job creation. Spierenburg and Brooks argue that, in practice, these approaches frequently end up 'disenfranchising local residents and reassigning their access and rights to the private sector'.[8] This is especially because, as I have pointed out, the farmers generally erect secure fences, so creating 'new spatial enclaves [. . .] leading to new forms of inclusion and exclusion, and with them new groups of "surplus people"'.[9] In terms of employment opportunities, proponents of hunting and game farming claim that more and/or better jobs are created by converting agricultural land to wildlife ranching. Spierenburg and Brooks counter this claim. They argue that research suggests wildlife farms are in fact less labour intensive, and also that, where jobs are indeed created, especially through tourism or hunting, they tend to be at a level beyond that at which locals residents can be employed: there may indeed be increased earning potential (fewer jobs, but better paid), but not to the benefit of local residents.[10]

Especially in areas like the Karoo, game farming has actually effected a movement of local residents from the farms themselves to informal settlements attached to small towns, Spierenburg and Brooks argue. They point out in this respect that it is profoundly ironic that 'the conversion to wildlife production in post-apartheid South Africa achieved what the apartheid government was never able to do – that is, to expel black labour tenants from white-owned farmland and create a racially purified space'.[11]

An article that Femke Brandt co-authored with Spierenburg, entitled 'Game Fences in the Karoo: Reconfiguring Spatial and Social Relations', offers an even more negative view of the social and economic effects of game farm conversion. Noting that '[p]rivate wildlife production – or game farming as it is known colloquially – has increased significantly in South Africa over the past few decades', these authors argue:

> In the Eastern Cape, more than 90% of mainly white commercial farmers and private landowners have diversified, adding or changing to game farming. About 7% of the commercial farms in the province have converted entirely to wildlife production. A high proportion of game farm revenues stem from hunting, especially trophy hunting, which accounts for 60%–80% of total income to the wildlife industry . . .[12]

They argue that in the process 'more farm dwellers have been displaced and evicted after the transition to democracy than during apartheid, with 2.35 million farm residents moving off farms between 1994 and 2004';[13] and that farm conversions to wildlife habitats, presented as ecologically driven, are frequently a tactic on the part of white farmers to avoid land claims and to consolidate private ownership, especially since game farming very often means more secure enclosure and fencing of land.[14] In the process, local communities suffer 'new forms of exclusion that threaten their livelihoods as well as their identities as they lose access to graves, land and livestock'.[15]

Brandt and Spierenburg also point out that the increase in land value effected by the conversion to game farming, referred to earlier, may negatively impact on land redistribution efforts, given the state's 'willing-seller-willing-buyer approach to redistribution'.[16] They claim that in 2011 agricultural land in the Eastern Cape could be bought for R8 000 per hectare, while 'in those areas most popular with game farmers', the price was up to R20 000 per hectare.[17] It is also extremely difficult and expensive to return a farm that has been converted to game ranching back to agriculture, as the conversion usually involves the removal or destruction of agricultural infrastructure.[18] While Brandt and Spierenburg have data that supports their argument that game conversion has displaced large numbers of people, and their argument has an ideological and symbolic logic to it, I am not necessarily convinced by the claim that such conversions are simply or necessarily cynical attempts by farmers to secure land ownership and displace farm dwellers. The political and economic history of farm ownership, including tangled narratives of (often undoubtedly paternalistic) systems of tenure, land usage and presence, suggest a more blurry picture. Reflecting on the fact that the conversion from agriculture to wildlife frequently involves the removal/destruction of existing farm infrastructure, for example, Brandt and Spierenburg say:

> Ironically, the 'wilderness landscape' thus buries a particular version of the past by 'rehabilitating' the land to create 'pristine' nature representing an era prior to colonial occupation. Settler descendants thus reoccupy and redesign the land by *erasing* the traces of processes of occupation and dispossession, as well as pre-colonial land-use practices. Stressing their contribution to nature conservation, they claim custodianship of this land for the good of all.[19]

It is the kind of argument that many scholars would make in engaging with societies marked by colonial histories, and suffers from the limitations frequently encountered in such arguments. They exhibit a level of historical and theoretical generalisation and abstraction that may make for elegant argument, but that do not engage the complexities and contradictions of the peoples, motives and histories they claim to describe. Is colonial and apartheid history erased in the destruction of farm infrastructure? For the most part, yes. Is that the intention of all the farmers? I very much doubt it. I suspect many simply do not want their springbok to have to run around barns and tractor sheds, which in any case no longer have any purpose.

In his comments on an earlier draft of this chapter, the anthropologist Knut Nustad reminded me that, while I may be correct in discounting the cynical intentionality part of such arguments, even with the best of 'intentions' on the part of farmers, the results of game conversions for the local communities may be the same. I take the point. My concern, though, is that arguments such as those of Brandt and Spierenburg may set up false oppositions that immediately demonise those involved in game conversions.

Also, one needs to distinguish between game farms, some of which are simply areas inhabited by game for hunting purposes with little or no infrastructure, and others in which elaborate lodges are built. As Njabulo S. Ndebele argues so elegantly in his essay 'Game Lodges and Leisure Colonialists' from his collection of essays, *Fine Lines from the Box: Further Thoughts about Our Country*, game lodges are almost inevitably a colonial presence upon the landscape, in which even the black South African visitor becomes a 'colonialist'. While acknowledging that a game lodge may be for 'moneyed white South Africans' a place where they can 'take refuge from the stresses of living in a black-run country', Ndebele sees the wilderness leisure space not as erasing colonial or apartheid histories, but creating a colonial present. And it is a present in which the black visitor experiences many of the symptoms of alienation identified by anti-colonial theorists such as Sartre, Fanon and Memmi in the colonised black subject: he or she experiences 'damning' or 'excruciating' 'ambiguities', and a 'stressful state of simmering revolt', as the identity that he or she lives, or which is forced upon him or her, is simultaneously denigrated and devalued.[20] He or she becomes a 'caricature of a tourist'. Ndebele points to the contradictions of the whole process of enclosing in the name of wilderness in that the

> fence reminds us that the game lodge, that inner core of cleared space, is not carved out of an unbroken wilderness but out of another contained space, sealed off from the country at large. It is a world of make-believe whose charm depends on the brief enjoyment it gives us to be a colonialist.[21]

We are on complicated ground here. In contrast to the simple definition of 'wilderness' set out by the US Wilderness Act, and the claims about pristine nature made in the tourist literature, notions of wilderness and wildness are always, perhaps necessarily and appropriately, stalked by contradiction, ambiguity and paradox, especially in their entanglement in South Africa with colonial and apartheid histories.

* * *

A friend who is aware that I am writing a book on wildness, himself an eminent scientist, phones me to tell me that the hottest item in the tourist trade in Fish Hoek is springbok skins dyed bright pink or green. The shop owner tells him that he cannot keep up with the demand, mainly from German tourists. The animals are dead, and the flesh no doubt consumed as biltong or venison. But we are both appalled at some fundamental level that seems to resist articulation by this phenomenon.

* * *

If wildlife farms or lodges can only exist in the absence of local peoples, we are into a system of absolutes that bears little relation to reality, and that assumes a notion of wildness that it can never achieve. Any tourist or trophy hunter who drives through the gates of an area enclosed by high game fences, often electrified, and assumes s/he is entering pristine wilderness is not thinking terribly hard. This is not to suggest that s/he will not encounter wildness: undoubtedly that will be there. Wildness can coexist with human habitation, if one thinks of it differently, and mixed mode farming, including game, can balance the needs of local peoples, tourism and ecology. Assuming that human influence negates wildness can, of course, also work conversely in tragic ways: in cases of tourists getting out of their cars in lion parks or game reserves to stroke or photograph the animals, on the assumption that they must be tame because they have been incarcerated by humans, with often fatal consequences.

Rewilding projects involving animals, especially predators, will almost always involve difficult decisions about how to manage either threats to human or animal life, or competition for resources such as grazing. Other projects, especially those in aquatic environments such as marine reserves or rewilded freshwater ecosystems, are less problematic. I have written elsewhere about the community trout project in the Eastern Cape, centring on Mnyameni Dam and the Cata River.[22] It was a project initiated with the local community by Martin Davies, an ichthyologist from Rhodes University. The Cata River had been stocked with trout around a century ago, and Mnyameni Dam has a self-sustaining population of trout that migrate up the feeder stream to breed during winter. Davies assisted the local community with a project to remove alien vegetation and restore the environment of the river and dam, and to set up a sport fishery. Income is generated by rod fees paid by visiting anglers, as well as payment of the guides who must accompany all who wish to fish. As well as some being employed as guides, other community members earn income by removing alien vegetation. It is an example of rewilding that seems to involve a sustainable natural resource, lightly managed and for the most part simply left to 'get on with it', which generates income for those who live in the region. I have since been informed that this apparently 'model' project is threatened by the discovery of a rare species of frog in the region, and consequent threats to remove the trout.

There is a community development project similar to that in Cata, at Thendela, on the edge of the Kamberg Nature Reserve in KwaZulu-Natal.[23] It is a village I frequently drove through on my way to fish at Kamberg in the years I lived in KwaZulu-Natal. Its state of impoverishment and underdevelopment, on the edge of a beautiful, well-managed nature reserve, always served as a harsh reminder of the apparent contradictions in the practices of conservation and rural development, even beyond the end of legislated apartheid. The Thendela Fly Fishing Project seeks to utilise the resource of the wild-spawning brown trout in the Mooi River, which runs through the village, to establish a trout fishery managed by the community. Fishing is by permit only, and is strictly on a catch-and-release, barbless hooks-only basis, so the resource is self-sustaining and rod tickets generate income for the community. The trout are of the Loch Leven variety from Scotland, having been stocked originally in 1890. The project is the result of collaboration between the KwaZulu-Natal Fly Fishing Association, in particular Richard and Linda Gorlei, and members of the Thendela community, notably Richard Khumalo, who now heads the project. Assistance and sponsorship have been provided by a number of members of the fly-fishing community to ensure that locals are trained as fishing guides (tuition was provided among others by leading Australian fly-fishing guide Peter Hayes), and that the associated commercial possibilities, including the tying and selling of flies courtesy of a donation of twelve 'Thendela training vices' by Jay Smit and instruction by Ian Cox, further serve to generate revenue for the community.

From childhood experiences of fashioning 'hooks' from fencing wire and 'line' from any suitable length of string, with which to catch fish on worms, grasshoppers and later rudimentary flies tied with feathers from the family's chickens, Khumalo is now himself an accomplished fly fisher and guide. As the river actually runs through the village, it is a visible reminder that conservation and human development are not mutually exclusive, that 'brown' and 'green' agendas need not necessarily be in conflict: there are chickens, vegetable patches, schools *and* wild trout.

* * *

A walk into the state forest, which abuts my colleague's home in Oslo, after a fresh snowfall reminds me of how quiet the wild can be. I am so used to noisy wilderness: the alternately melodic or cacophonous sounds of the coastal bush in KwaZulu-Natal; the gushing of the river and the far-off whistling calls of raptors, or the indignant quacking of a pair of yellow-bills startled as one rounds a sharp river bend, fishing upland trout rivers; the insistent Piet-my-vrou, and the hadedas voicing their complaint to all. But in this forest it is utterly still. I had read of the quietness following snow, but never experienced it. The only sounds are the puff as powdery snow drops from the conifers, and the gentle gurgle of a tiny stream running mostly under the snow and fed by snowmelt from the sun, which has unexpectedly made an appearance. The occasional bird calls mutedly and infrequently.

And then there is the almost silent hiss of the skiers intermittently and unexpectedly appearing at my back, descending the track at a speed that will leave me impaled like a barn rat on a pitchfork if I don't watch myself.

*   *   *

Part of the problem with many conservation efforts or projects is that they seem to assume, especially in their emphasis on 'restoring nature', that the land in question has no political history; or they attempt to conceal this history under the 'nature' argument. I want to explore ways of thinking differently about conservation, land and people, which may avoid some of the impasses. Knut G. Nustad's book, *Creating Africas: Struggles over Nature, Conservation and Land*, is enlightening in this regard. His argument is largely about 'nature reserves' rather than rewilded areas, though the distinction between the two is often blurry, and also his comments frequently apply to both contexts. Nustad's crucial point is that the founding principle on which ecological damage is diagnosed and remedial action proposed involves a fallacious assumption about the division between 'nature' on the one hand and 'society' on the other. Unless we can think and act beyond this dualism, he argues – correctly in my opinion – we will always face impossible contradictions and multiple dead ends in our environmental action.

Nustad says that '[m]any in the West think of nature conservation as something that is inherently good'; that it is about 'protecting nature, about looking after those remnants that humanity has not already destroyed through extraction, industrialisation, urbanisation and pollution'.[24] In this model, humans are regarded as having negatively impacted on nature, and so the solution is to set aside 'nature reserves' that will be protected from human influence. The 'nature' versus 'society' division is starkly evident in this thinking, and Nustad points out that '[p]rotected areas have a special appeal to some of us because we think of them as separate from society'.[25] In visiting such areas, we are 'getting away from it all'. In fact, he states, the International Union for Conservation of Nature uses a grading system for wilderness areas according to 'how much human intervention in nature' has occurred, with the highest level of protection afforded to those with the least human influence.[26] And yet, Nustad argues, ironically, we are attempting in so many of our conservation efforts to solve a problem by the same means that caused it.[27] A sense that human society was somehow separate from, and often superior to, nature has served as legitimation for the systematic plundering and despoliation of the environment; and yet now we replicate that distinction in apparently more benign ways by setting apart 'nature reserves'. In so doing, in prizing 'nature' over 'society' in these areas, we almost inevitably generate conflicts with the local communities whose lives and livelihoods are affected by removal from the land, or proclamation of the land as off limits.

In terms of the goal set by some conservationists that at least 10 per cent of the globe should be placed 'under protection', Nustad points out that we had already reached this goal by the late 1980s, with 'over 105 000 protected areas in the world, covering more than 20 million square kilometres'.[28] It is in one sense good news; in another, it suggests just how pervasive the dualist model actually is. We are 'protecting' these 'nature' areas. Does that discharge our environmental obligations? What are we doing to the rest?

Nustad argues, in contrast, for an approach in which 'humans, animals and their environments are treated as mutually constitutive, and as constituting different realities in the making'; one 'that seeks to transcend oppositions between humans, environments, animals and objects'.[29] The dualist approach in South Africa and many other places has its roots, he claims, both in 'a gradual elitist restriction on access to animal and plant resources' and in 'nationalism expressed as the romantic celebration of pristine nature as a mirror of national character'.[30] On the latter point, we should note the naturalisation and glorification of apartheid violence in attributing the names of wild animals, frequently in Afrikaans, to military vehicles, weapons and aircraft developed in South Africa in the time of the arms embargo: we had the *buffel, rooikat, olifant, rooivalk, ratel,* hippo, nyala, mamba, *renoster, luiperd,* bateleur, and so on.

The idea of a nature apart from society in Western thought may have many roots and branches, including Greek philosophy, Judaeo-Christian theology (some may say a misreading thereof), Enlightenment rationalism, Cartesian philosophy, and a whole range of others, including the Disney industry. But Nustad is particularly interested – in the case of South African environmental debates and those in other countries with colonial histories – in the era of industrialisation and urbanisation, in which the despoilation of landscapes and the exposure of people to frequently brutish and impoverished urban living conditions engendered in many 'a romantic longing for regions that had not been transformed by these changes, a nature pure and pristine'.[31] Frequently idealised representations of natural landscapes in paintings or literature offered an alternative to the harsh realities of rapidly advancing industrialised capitalism. He also explores the association in Europe, and in colonies, of hunting with the political elite, in particular with notions of masculinity, and hence the importance of access to areas in which these activities could be pursued. As animal numbers declined, the politically powerful sought to restrict access to areas set aside for them to exercise their 'hunting rights'. 'Hunting' became transformed over time into a sporting code (the Hunt, not just 'hunting'), with its own self-imposed restrictions (only shooting game birds 'on the wing', for example, or in the case of fishing, say, only using flies, not bait or lures, in pursuing trout), though the worldwide demand for animal products meant there remained an economic rationale for hunting, and game meat was also an important food source, especially in the colonies.[32] The political elites frequently banned other forms of hunting or fishing, and imposed heavy fines on 'poaching'.[33]

Along with this, there was an increasing interest in natural history, fuelled by debates about Darwin's work on evolution, as well as, I assume, the encounters with a wide range of plant and animal species in or from the colonies unfamiliar to residents of Europe, and the associated development of museums of natural history eager for specimens.[34] This interest 'soon began to extend beyond the elites and was taken up by the middle and working classes', but the 'connection with hunting and the hunting ethos remained: many hunters collected for museums, while the museums tended to be controlled by the hunting elite'.[35]

A loose alliance of elite hunting interests and natural history/ preservation interests seems to have been at play in the declaration of public parks in South Africa in the early twentieth century, but both seem to have reinforced the dualistic model. Nustad says in this regard that '[t]he Hunt and the idea of national parks were based on a new meaning being given to nature and wilderness with the advent of industrialisation, albeit through different routes. But they both helped contribute to the development of a dualis[m] [. . .] that created a split between the world of humans and the world of nature.'[36] In her study on the Kruger National Park, Jane Carruthers refers to the racial politics involved in the park's creation, which involved 'white perceptions that Africans destroyed vast numbers of wildlife, that they trespassed in order to do so, that they killed in a cruel manner, that they spoilt the recreation of sportsmen and managed to evade wage labour by subsisting on wildlife'.[37] As Elsie Cloete remarks, such assumptions are deeply ironic in view of the fact that the Kruger National Park and the bulk of such parks in southern and eastern Africa 'were established principally to offset the profligate slaughter of animals by white-skinned hunters'.[38]

The model for such parks, which involved moving people out of the proclaimed areas and erecting fences, is referred to as 'fortress conservation'.[39] Fortresses are, of course, designed to keep some people in and some people out, or in this case, animals in and people out, or nature in and society out. Quoting Agun Agrawal and Kent Redford, Nustad says that between eight million and 130 million people worldwide have been relocated as part of conservation projects.[40] The variance between those figures is so considerable as to raise questions about their validity, and that concern is exacerbated if one takes account of the 2.35 million people Brandt and Spierenburg claim have moved off farms between 1994 and 2004 in South Africa alone, cited earlier. Whatever the actual figures, there is no disputing the fact that large numbers of people have been affected, and Nustad quotes Mark Dowie's comment that 'tribal leaders on almost every continent' identify as 'culture-wrecking institutions' Conservation International, the Nature Conservancy, the World Wildlife Fund and the Wildlife Conservation Society, alongside the more usual suspects in Shell, Texaco, Freeport and Bechtel.[41]

Earlier in this chapter I discussed the Community-Based Natural Resource Management approach, which was developed to try to balance the needs of local communities with conservation imperatives. Such projects frequently fail because they are based on 'oversimplified understandings of both poverty and biodiversity',[42] and also, as Paige West notes, because it is often not clear 'whether Community Conservation Programmes are conceived of as conservation programmes with development benefits, or as development programmes beneficial for conservation'.[43] There is also a kind of built-in duplicity in many such programmes in that in participatory or bottom-up approaches to conservation projects, 'the goal that is supposedly to grow out of the process of development itself is actually, of necessity, defined prior to the development intervention',[44] including goals defined by conservation bodies with funding for specific projects (and not others). Imposing solutions developed elsewhere, even indirectly, rather than working with communities on their own sets of solutions is one of the most significant pitfalls in development work; as is assuming that communities will 'buy into' agendas or priorities set by funding organisations, and not by those living on the ground.

How then does one find a way through or beyond the dualist approach? The key is to understand humans, animals and plants as mutually influencing and dependent organisms within the same world. It is an approach that resonates strongly with the ideas of Snyder and Monbiot, which provide a larger frame for my argument in this book. Nustad makes the point by invoking the ideas of the anthropologist Tim Ingold:

[I]nstead of conceiving of human beings as existing in a social world inhabited solely by other humans, with animals and plants existing in a natural world inhabited solely by other non-humans, we should conceive of ourselves as organisms-in-environments, environments inhabited by both humans and non-humans alike [. . .] Organisms, including humans, are both formed by the environment in which they live and in turn form that environment; it is meaningless to speak of environments and organisms as separate entities. This, as [Ingold] points out in later writings drawing on the work of Deleuze, also means that we need to treat environments not as already constituted entities with which organisms interact, but as movement or lines of movement, as domains of entanglements rather than environments . . .[45]

It is an approach – echoing Snyder's *practice* of the wild – that emphasises not just ways of 'knowing' the world, but ways of 'being' in the world, with the development of the requisite 'skills'.

The point is that organisms and their environments are mutually constitutive, and Ingold initially refers to this as a 'dwelling perspective'.[46] He later apparently indicated his regret in having chosen this term, because 'of its connotations of harmony and integration', noting that the 'dwelling of one species-in-environment may mean the destruction of another species-environment nexus'.[47] To me, the ambiguities of dwelling are precisely the point. A rather wise person once pointed out that a house divided against itself cannot stand, and as a species humans have been selfish and greedy dwellers, to the detriment of the home and all of its occupants.

Can we dwell differently? Indeed in conservation terms, can human dwelling be accommodated in areas in which biodiversity and rewilding are promoted? Or to ask the question differently: can wildness and humans coexist? My answer is yes, because we are not in a game of absolutes.

To take an example, within the Kruger National Park, there are many camps in which visitors can stay overnight. All are protected by fencing of varying strengths. The camps range from fairly rudimentary places to pitch a tent or park a caravan, to something resembling a small town, with a supermarket, Wimpy Bar and Mugg and Bean coffee shop. There is even a nine-hole golf course near Skukuza. If you read the accounts of visitors in a collection like *101 Kruger Tales: Extraordinary Stories from Ordinary Visitors to the Kruger National Park*, for example, the sense of living in the wild in a (mostly) protected space is strongly evident, as is the sense that, despite the massive volume of car traffic in the park, one is still encountering 'wild animals' in their 'wild habitat'. Accounts of seeing kills, of animals attacking cars, animals circling camps or in some cases entering them, are narrated in a breath of awe at what one is encountering. Rightly so. Some of the animals are no doubt habituated to cars, which despite isolated incidents they do not apparently associate with danger, as they do in areas that are frequently hunted. But I would invite anyone questioning their 'wildness' to step out of their car to see. A visitor to the Park, Kathleen Stevens, makes the interesting observation in *101 Kruger Tales*:

> It is strange how I would never entertain stopping my car in the middle of Kruger, casually getting out and wandering around on foot near a waterhole – yet this is exactly what we all do when we arrive at any of the hides dotted around the Park. I suppose the presence of a man-made structure and a neat clearing for parking makes the place seem less untamed. But why should it? After all, if we humans enjoy watching game amble down to drink from the shade of a hide, it stands to reason that predators would, too.[48]

It should also be noted that, as the editor of that volume, Jeff Gordon, points out, even if you drove every single road of the Kruger Park, with a clear view of 100 metres to each side of the vehicle (which is virtually unheard of), you would only see 2.5 per cent of the park.

It is not Wimpy Bars *or* leopards. It is not humans *or* nature, wild *or* civilised. It is always, necessarily, both. As the example of the Kruger camps suggests, I am not proposing putting people back into contact with dangerous predators; there are many qualities or gradations of wildness. Snyder captures the possibilities of wildness from the domestic to the backwoods:

> Every region has its wilderness. There is the fire in the kitchen, and there is the place less traveled. In most settled regions there used to be some combination of prime agricultural land, orchard and vine land, rough pasturage, woodlot, forest, and desert or mountain 'waste'. The de facto wilderness was the extreme backcountry part of that. The parts less visited are 'where the bears are'. The wilderness is within walking distance – it may be three days or it may be ten. It is at the far high rough end, or the deep forest and swamp end, of the territory where most of you all live and work. People will go there for mountain herbs, for the trapline, for solitude. They live between the poles of home and their own wild places.[49]

It is a point made differently by Arthur Carhart, who argues that there are in fact 'several gradations from the absolute wilderness, toward the semi-urban picnic zone of our wildlands, that can deliver to many people approximately the full impact of the absolute wilderness'. This notion of wildness does not exclude human changes to landscape, for these 'zones in wildland classifications may have in them old wagon roads, dilapidated sawmill structures, abandoned mines, even fresh jeep tracks and still supply many people a true wilderness experience'.[50]

It is intriguing that in maritime environments, shipwrecks or artificial reefs created by dumping tyres or rubble, very clear evidence of human intervention, are highly valued by divers and fishermen, in that they create complex ecosystems that support a wide range of species. They become sought-after sites for viewing marine life or catching fish, crayfish or crabs. And as anyone will testify who has been buzzed by the large sharks that wrecks seem to attract, there is nothing tame about them.

To return to the issues raised earlier in this chapter about the conversion of farms to game ranching, I would add that 'old wagon roads', 'fresh jeep tracks' and 'dilapidated saw mills' can also be found in strong, thriving ecosystems, and that consequently, if we dispense with the misconception that 'wild' or 'nature' necessarily mean human absence, rewilding and conservation need not necessarily occur at the expense of local communities and their livelihoods.

We might then start to think more imaginatively and creatively about how we live in the world. Jack Turner says Henry David Thoreau expressed the possibilities so well, when he talked of 'wild nature, wild men, wild dreams, wild house cats, and wild literature', which he associated with the 'the good, the holy, the free . . . [i]ndeed . . . with life itself'.[51]

* * *

One of my own signs of wildness is a barbed wire fence – probably the cheapest and most rudimentary form of fencing you can erect, and as good as nothing in an urban environment. With a little practice they are easy to get through, though they threaten you with tetanus if you're unwary, as the scrags of animal hair on the barbs suggest. For the Old Testament prophets, the rewilding of cities laid bare by war was indicated by the presence of owls, and I have probably seen more owls on fence posts driving home in the evenings after a day's fishing than anywhere else. Barbed wire fences sing in the wind. They tell you that you are away from the world of traffic and tax returns, that you can feel like a kid again. That here your identity as a lawyer or plumber or school teacher counts for very little; what counts is the strength of your legs, the sharpness of your perception, and the humility of your soul.

* * *

I want to conclude with a personal account of an aspect of wildness/ conservation practice that I find more problematic, though I suspect it will become more prevalent with the lucrativeness of tourist foreign currencies: what might be called 'safari camps'.

A recent overnight breakaway with the administrative staff from my faculty office took us to what is advertised as a 'Private Game Reserve', offering the opportunity to see 'the most free-roaming safari animals under two hours from Cape Town'. The promotional material for the resort advertises it as a 'Big 5 Safari'. It is arguably the largest of a number of similar 'camps' in the Western Cape, not far from Touws River.

After a leisurely lunch, and an hour or two of planning work for the new semester, we headed out for a game drive, which is one of the resort's main attractions. On the way down from my room in the main accommodation building, I noticed two elephants walking across the plain with apparent purpose. We departed for the drive with four or five other vehicles, each carrying around twenty or so tourists, and as we entered the gates into the reserve itself, we encountered the two elephants standing close by in the Karoo scrub, eating enthusiastically from the ground, and apparently unmoved by the numbers of game vehicles jostling for position around them. I was a little suspicious of what they were eating so assiduously in this spot, which seemed no different from the rest of the landscape. Then our driver moved the vehicle to give us a different angle for photographs, and the answer was revealed. A bakkie load of carrots had been deposited there, conveniently close to the main gate and within the focal length of most cellphone cameras from the road.

The rest of the drive played out in similar fashion. The hippos in the nearby dam seemed to be doing their own thing, though I wondered what they ate in this dry scrubland; but the zebras, wildebeest and rhinos were all easy to find by simple virtue of their congregating at the sites at which they were fed. The drivers of the vehicles seemed to vie with each other in getting their guests into the best spots for photographs, sometimes unwisely placing the vehicle in the path of a moving animal, or, in the case of the lions (also fed), by parking too close to an animal, which was then penned in next to a fence. We were informed by our guide that the lions had been rescued from a 'canned lion' hunting business, and that the male lions have been sterilised, and so cannot breed. For those who had never seen a live elephant, rhino or wildebeest, it was no doubt a wonderful experience; but to all intents and purposes we were really in a large, open-air zoo.

When I questioned our driver about the feeding of the animals, he confirmed that the area does not have the vegetation to sustain the population of animals that inhabits it, so feeding is the only alternative. (I heard later that the resort had previously been in trouble with conservation authorities for keeping more animals than the land could sustain, but have not been able to corroborate that.) Guests entering the resort are required to sign indemnity forms against personal injury, and notices in the resort warn against approaching animals on the mistaken assumption that they are tame (correctly, in my opinion: an animal habituated to a game drive vehicle may look rather more aggressively or hungrily at a lone bipedal mammal wandering into his/her space). What the resort offers is the sight and presence of animals, not the experience of encountering animals in environments in which they can live out anything approximating a 'natural' life. In its defence, the resort does not promise an experience of 'wild' animals (although its promise of the 'Big 5' may be stretching the truth a little). It emphasises its role in rhino conservation and the education of the public, as well as its commitment to balancing 'brown' and 'green' agendas in relation to the sustainability of its operations: its involvement in providing employment and livelihood opportunities for members of the surrounding community; and its commitment to environmentally sound practices in managing water, heating, and so on.

The large numbers of guests who frequent the resort seem to confirm its attractiveness for many, but it is really a large-scale farming operation rather than a rewilding project. In saying that, I want to avoid what might too easily become either prescriptiveness or patronising judgementalism in describing the practice of 'exhibiting' animals in such ways. Nor do I want to diminish the experiences of tourists who may never otherwise have a chance to see a lion in the flesh (if not in the wild). But the question is for me, who benefits and what is at stake? The animals appear well fed, and are apparently given veterinary treatment when ill. Clearly the owners are on a lucrative wicket. The visitors seemed generally content, and employment opportunities have been created in a landscape otherwise comparatively inhospitable for any kind of farming besides sheep.

What the long-term effects on the environment of such a large population of animals, even with supplementary feeding, will be, remains to be seen, as will the environmental costs of sourcing and transporting food (one elephant alone consumes approximately 200 kilograms of food daily).

Perhaps more fundamentally, though, rewilding is about sustainable systems, involving little human intervention, in which ecosystems can regenerate, and in which human presence is welcome, but not necessary, and should not be intrusive or controlling. In contrast, the 'safari park' in which I stayed was about placing animals in view of people in sufficient proximity and numbers to please even the most demanding customer, with little or no discomfort to the tourists, and at a venue close enough to a major tourist city not to disrupt the itinerary of other major attractions in the Western Cape (wine tasting, visiting Robben Island, travelling to Cape Point, and so on). While the animals are ostensibly the attraction, actually the entire experience is centred on people, and their ability to purchase (mostly with foreign currency) an appearance and performance by the animals. Perhaps that is why I felt more offended than I expected at the sight of elephants eating carrots in Karoo scrubland.

4

Wildness and Local Language

Smeuse is an English dialect noun for 'the gap in the base of the hedge made by the regular passage of a small animal'; now I know the word smeuse, I notice these signs of creaturely commute more often.
— Robert Macfarlane, 'The Word-Hoard'

It's a city perched on the edge of water, a city of two rivers, blown by every wind named and unnamed, by ban-gull and haugull, blinter and flist.
— Esther Woolfson, *Field Notes from a Hidden City*

The sea speaks a language polite people can never repeat. It is a colossal scavenger slang and has no respect.
— Carl Sandburg, *The Complete Poems of Carl Sandburg*

I have written elsewhere about tracking as a form of reading, requiring the sophisticated interpretation of signs – spoor, blood smear, scat, broken branches, displaced stones, flattened grass, traces of hair or feathers, and so on – to locate animals or understand their behaviours, drawing especially on the work of Louis Liebenberg.[1] Rewilding involves expanding our ability to read the signs of the systems and life forms around us: wind, clouds, rain, tracks, waterways, nests, droppings, waves, currents, movements, behaviours, colouration, texture, shape, and myriad identifiers or differentiators. It is the work of several lifetimes. But there is a qualitative difference to such a life; and thankfully we do not have to start from scratch: there is an enormous body of knowledge from which to draw, and the languages to capture it, should we choose to find it. In this chapter, I look specifically at the possibilities of local languages, or languages of locality, in this regard. Human language is probably the largest, most complex, diverse and contested subject that anyone can engage, so I need to make it clear that my aim in this chapter is necessarily modest and partial: to explore some examples of particularly interesting engagements of (mostly) local languages with the wilding or dewilding of their environments, and in the process themselves. The examples are deliberately disparate in terms of geographical location and historical period, to suggest the possibilities of languages of (re)wilding in divergent contexts.

*   *   *

The river in its mild flood was impressive. It had run through the mile-long canyon, turbulently over its bed of boulders and against the steep rock sides, and now it came from the narrow mouth of the canyon, flurried white between the tall gray walls by a ledge of rock that ran out from the far bank. In the moment of sudden release it spread into a great wide pool, the white of its hurry lost in a current-creased surface. There were deep eddies on either side, under the lee of the rock walls, but the river slid on between them, forcing them apart, confining them above itself, until it had spread to a fan at the tail of the pool and claimed the whole wide bed for its flow. Then it was broken and white again in the long rapid that hurried down to the next pool. — Roderick Haig-Brown, Return to the River

*   *   *

Scotland is known for many great things, not least among them its whiskies, but not for its fine weather. Anyone reading Ian Rankin's

Detective Inspector Rebus series of novels, for example, which are set for the most part in Edinburgh, will note the range of words or expressions used to differentiate qualities and characteristics within the category of what would probably be called 'rain' by most non-Scots. Chris Robinson and Eileen Finlayson's book *Scottish Weather* is a compendium of terms or expressions used by Scottish speakers of English to describe weather (mostly bad), and in some cases, by metaphorical extension, human behaviours. Some notable examples are:

blatter: to rattle, beat violently (often used of rain or hail); a violent rain or hailstorm. A period of sunshine during unsettled weather, or a moment of joy in troubled times may pessimistically be called the 'blink afore the blatter'.

dreep: a light, steady fall of rain.

dreich: dull, dreary, gloomy weather.

feechie: foul, dirty, rainy and puddly. Feechie can be used for anything messy or unpleasant. Feech! is an exclamation of disgust.

onding: a heavy fall of rain or snow, a downpour. Ding is a verb meaning to beat or strike with heavy blows. So a real onding of rain has some real force behind it. Onding can also be used figuratively to mean an assault, attack, onset, outburst of noise, talk, etc.[2]

While the extent of such language may be daunting for non-Scots, the point is in fact not to obscure, but to distinguish or elucidate; to provide ways of describing the multiple manifestations of Scottish weather.

In similar vein, fly fishers have a whole vocabulary to describe the characteristics of the streams they fish, which includes terms such as 'runs', 'riffles', 'glides', 'pockets', 'roils', 'braids', 'seams', 'tails' and 'scours' (as well as the more obvious 'undercut banks', 'eddies', 'backwaters', and so on). Again, the intention is not to be arcane or exclusive, but rather – for a stretch of river that a non-fly fisher would probably differentiate only into 'rapids' and 'pools' – to identify accurately the specific types of water that one approaches and fishes differently.

Anyone reading George Monbiot's *Feral* will be struck by his routine use of words such as 'carr', 'fridd' or 'brae' in describing landscape.[3] These would probably be unknown to most readers, but Monbiot uses them to differentiate particular features of the landscape because they are appropriate. He is, in an important sense, 'rewilding language', not for its own sake, but in seeking nuanced and intimate ways of naming and apprehending the environments in which we live: a language that can register the kind of 'dwelling perspective' suggested by Tim Ingold in the previous chapter. Language is, of course, fundamental to our apprehension, categorisation and valuation of the worlds in which we live, and in this chapter I want to consider the possibilities and implications of 'thinking wild' for language.

*   *   *

The oystercatchers call a lovely, hasty, crazy, frantic call, trailing it wildly across the skies. They are commonly regarded as Gaelic speakers, with their cry of 'Bi Glic! Bi Glic!', 'Be wise! Be wise!' — *Esther Woolfson,* Field Notes from a Hidden City

*   *   *

The environmental writer Robert Macfarlane has for many years been concerned with collecting unusual words for landscapes, plants, animals or weather conditions. In this regard, he recalls being given 'an extraordinary document' in the coastal township of Shawbost on the Outer Hebridean island of Lewis. It was entitled 'Some Lewis Moorland Terms: A Peat Glossary', and included 120 Gaelic words or terms for the local landscape.[4] Some examples that catch Macfarlane's imagination are:

coachan: a slender moor-stream obscured by vegetation such that it is virtually hidden from sight.

feadan: a small stream running from a moorland loch.

fèith:	a fine, vein-like watercourse running through peat, often dry in the summer.
rionnach maoim:	the shadows cast on the moorland by clouds moving across the sky on a bright and windy day.
èit:	the practice of placing quartz stones in streams so that they sparkle in moonlight and thereby attract salmon to them in the late summer and autumn.
teine biorach:	the flame or will-o'-the-wisp that runs on top of heather when the moor burns during the summer.[5]

In these and many other examples cited in this chapter, the ability of a single word to capture a feature of landscape, concept, practice or visual image that requires definition in a sentence or even a paragraph in another language is extraordinary. As Macfarlane comments, 'Words are grained into our landscapes, and landscapes grained into our words.'[6]

At around the same time that the 'Peat Glossary' suggested a range of additional terms to Macfarlane, the *Oxford Junior Dictionary* was doing the opposite. Oxford University Press admitted that it had deleted from the new edition of the dictionary a range of words that 'it no longer felt to be relevant to a modern-day childhood', including 'acorn, adder, ash, beech, bluebell, buttercup, catkin, conker, cowslip, cygnet, dandelion, fern, hazel, heather, heron, ivy, kingfisher, lark, mistletoe, nectar, newt, otter, pasture and willow'. In their stead had been added, 'attachment, block-graph, blog, broadband, bullet-point, celebrity, chatroom, committee, cut-and-paste, MP3 player and voice-mail'.[7] (That was about eight years ago, and I wonder now whether anyone still uses MP3 players, though I hope the otters, kingfishers and herons will be around for millennia to come.) Macfarlane laments:

[I]t is clear that we increasingly make do with an impoverished language for landscape. A place literacy is leaving us. A language in common, a language of the commons, is declining.

Nuance is evaporating from everyday usage, burned off by capital and apathy. [. . .] The terrain beyond the city fringe is chiefly understood in terms of large generic units ('field', 'hill', 'valley', 'wood'). It has become a blandscape. We are *blasé*, in the sense that Georg Simmel used that word in 1903, meaning 'indifferent to the distinction between things'.[8]

It is an impoverishment not limited to English, he notes, but evident also in Irish and Gaelic. He does, however, report substantial public outcry to the dictionary deletions, and several projects to counter such moves, including that by Monbiot himself.[9]

In response to the sentiments expressed above, as well as the resources of the 'Glossary' and the substitutions of the *Dictionary*, Macfarlane set out on his own project of collecting terms for place and landscape, which eventually appeared as the book *Landmarks*, comprising nine individual glossaries with commentary on particular writers who have significantly engaged place and environment in their work. He notes that the words are those used by people who read the landscape, mainly for work, but also for leisure: crofters, fishermen, farmers, sailors, scientists, miners, climbers, soldiers, shepherds, poets, walkers and the like.[10] In this regard, he says, 'I became fascinated by those scalpel-sharp words that are untranslatable without remainder. The need for precise discrimination of this kind has occurred most often where landscape is the venue of work.'[11]

The terms and their definitions are often evocative and delightful. 'Ammil' is a 'Devon term for the thin film of ice that lacquers all leaves, twigs and grass blades when a freeze follows a partial thaw'. 'Pirr' is a Shetlandic word meaning 'a light breath of wind, such as will make a cat's paw on the water'. 'Zwer' is the onomatopoeic word for 'the sound of a covey of partridges taking flight' in Exmoor, and 'crizzle' in Northamptonshire dialect describes the freezing of water, evoking 'the sound of a natural activity too slow for human hearing to detect'.[12] Others are, as Macfarlane reminds us, 'ripely rude'. The West Country term for 'a very substantial cowpat' is 'turdstool'. 'Ujller' is the Shetlandic term for the 'unctuous filth that runs from a dunghill'. And should Gerard Manley Hopkins have wanted an alternative dialect term for a kestrel in the title of his poem 'Windhover', he could have opted for 'wind-fucker'.[13]

Macfarlane's accounts of the extraordinary intricacies of language, place and identity in the work of particular writers are for me especially compelling. Norman MacCaig's plea in his Luskentyre poem – 'Scholars, I plead with you, / Where are your dictionaries of the wind . . .?' – in fact proves to be a spur for the entire project.[14] Tim Robinson, who spent 40 years writing, painting and mapping the west of Ireland, remarks that 'the landscape speaks . . . Irish'; and he refers to the 'language we breathe' as providing 'our frontage onto the natural world'.[15]

If all of this sounds overly Eurocentric, Macfarlane refers to a project that he encountered that is similar in aim to his, but massively more expansive in scope, by a scholar living in Qatar. Abdal Hamid Fitzwilliam-Hall has been working on a study, which at the time Macfarlane encountered it, included 140 languages, contained '50 000 separate terms or headwords' and spanned 3 500 pages.[16] It is apparently still in progress . . .

I need to add a caveat here, and one of which Macfarlane was aware, as indicated by his inclusion of current urban neologisms. Drawing attention to words or phrases that are rapidly being lost is not to advocate that we all adopt a quaint form of antiquated speech, the sort of archaic, bucolic language for which the Irish poet Seamus Heaney's early poems were unjustly caricatured. It is simply to recognise that with the loss of such terms, we are losing our ability to read and understand the environments in which we live, and the biological systems on which we depend for our sustenance, however far along the food chain we might think we are. Whether one calls it a *fèith* or not, for example, it is important to notice whether the moorland stream, which may dry up in summer, is now always dry, and what that means; or to register that no one bothers with *èit* any more in the local river, as the salmon runs have stopped.

A term coined for weather more recent than some of Macfarlane's examples came from two Australian scientists studying the smells of wet weather. In an article in the influential journal *Nature* (7 March 1964), Isabel Bear and Richard Thomas introduced the term 'petrichor' to describe the evocative smell of rain falling on warm ground.[17] The term is derived from the Greek words 'petra' (stone) and 'ichor' (the blood of the gods in Greek mythology).[18] The pair of scientists discovered that the smell was the result of moisture entering porous rocks and causing miniscule amounts of yellowish oil to be released, which produces the distinctive smell. The wind, which often accompanies rain, helps to disseminate the smell.[19] We are smelling the oil from the stone, or the 'blood of the earth'. It is a smell deeply familiar to almost all of us, and apparently causes cattle in drought-stricken areas to become restless, presumably in anticipation.[20]

* * *

I will arise and go now, for always night and day
I hear lake water lapping with low sounds by the shore;
While I stand on the roadway, or on the pavements grey,
I hear it in the deep heart's core.
(William Butler Yeats, 'The Lake Isle of Innisfree')

* * *

Language does not only contribute to apprehensions or misapprehensions of landscape. As the literary scholar Leon de Kock has argued, it may also be fundamental to attempts to change patterns of land ownership and usage, and instrumental in the processes of dispossession. It can 'dewild' or 'reculture' it. Nineteenth-century missionaries in the Eastern Cape were at pains to rename and reconceptualise African relationships with the land as part of the 'civilising' project of Christianity. As De Kock points out, the initial model was that of the 'civilised haven' of the mission station surrounded by the 'savage terrain' around it.

But with the expansion of mission activities and presence, the large mission stations became educational, spiritual and economic centres of their own, domesticating the surrounding regions.[21] Fundamental to this process was the 'dewilding' of landscape and the language used to describe it, and the regulation, settling, cultivation and mapping of terrain to create 'orderly communities'. As the anthropologist Jean Comaroff has argued, 'agrarian metaphors came to pervade the evangelists' vision of a Christianised Africa: it was a "wilderness" to be turned into a "fruitful field"'.[22] The missionary W.R. Thompson said in his report to the Glasgow Missionary Society in 1827, for example:

> It is a comfort to me that I can shew brickmakers, thatchers, sawyers, ploughmen and jobbers at ditching, hedging and field work, who do wonderfully well considering the master they had to instruct them. Where formerly a wilderness of long grass was, and the soil never turned up since the Flood, we now have growing many of the necessaries, and even some of the luxuries, of life. A neat little village has been formed, inhabited by those who a little while ago roamed the world at large, as wild and savage as their old neighbours, the lions and tigers of the forest . . . If you except the black faces, a stranger would almost think he had dropped into a little Scotch village.[23]

While the approach of the missionaries might have been rather more ameliorative, at least on the surface, colonial authorities were more aggressive and bellicose in asserting control of language and subjects. Sir Harry Smith illustrates this in his statement to the defeated Xhosa after the War of the Axe (1846–7), in rhetoric that echoes the creation narrative in Genesis:

> Your land shall be marked out and marks planted that you may all know it. It shall be divided into counties, towns and villages, bearing English names. You shall all learn to speak English at the schools which I shall establish for you . . . You may no

longer be naked and wicked barbarians, which you shall ever be unless you labour and become industrious. You shall be taught to plough; and the Commissary shall buy of you. You shall have traders, and you must teach your people to bring gum, timber, hides etc. to sell, that you may learn the art of money, and buy for yourselves.[24]

In such contexts, English became a means to shape understandings of human relations with landscape along European lines. It became complicit with, even instrumental in, initiating a process of land dispossession, which was to culminate in the notorious Land Act of 1913, through which the Union Parliament allocated 87 per cent of South Africa to white ownership.

* * *

It was a matter of chance that I should have rented a house in one of the strangest communities in North America. It was on that slender riotous island which extends itself due east of New York and where there are, among other natural curiosities, two unusual formations of land. Twenty miles from the city a pair of enormous eggs, identical in contour and separated only by a courtesy bay, jut out into the most domesticated body of salt water in the Western Hemisphere, the great wet barnyard of Long Island Sound.
— *F. Scott Fitzgerald,* The Great Gatsby

* * *

As the statements quoted above from W.R. Thompson and Sir Harry Smith suggest, the landscape named and fashioned by English is modelled on that of the island in which that language originated. What it cannot name or apprehend is dismissed as 'wildness' that must be suppressed. English, in fact, turns out to be a rather poor language for engaging with African landscapes. The history of South African literature in English reveals many examples of the ways in which authors have turned to Afrikaans and appropriated its terms for the

South African landscape – *veld, koppie, berg, kloof, donga* (via isiXhosa), *spruit,* to name a few. In many cases, these usages reflected attempts to write more 'authentically' in South African voice, and Afrikaans gave name to the movements, titles or journals that engaged in non-English appropriations: *Voorslag, Donga, Sjambok,* the Veldsingers, *Trek,* or the *Purple Renoster.* That, however, is the subject of another, rather different, book.

But it is certainly the case that Afrikaans has provided a rich resource of terms for speaking about the South African landscape in ways that English apparently does not. Some of the better-known terms are listed below. Part of the difficulty in engaging with such terms, though, is that their dictionary definitions seek English-word equivalents, which means that the idiomatic fullness of the term is frequently lost. To illustrate the point from another language, isiXhosa, *ukutshotshobela* is probably best translated as to 'wriggle/squirm closer to the fire on your bum', although Kropf in his famous 1915 dictionary rather prudishly defines it as 'to draw near (to the fire)'.[25] To return to Afrikaans, however, in the first example below, for instance, defining a 'kloof' as a 'gulch' seems particularly unhelpful. In such cases, I have sought where possible and appropriate to offer slightly more expansive and suggestive definitions. It is also striking how many Afrikaans words have entered South African English (and in the case of 'trek', global Englishes), so that the same word is offered by dictionaries as the Afrikaans term and English translation (*kloof* – kloof; *veld* – veld; and so on).[26]

> *Kloof*: kloof, gulch, ravine or gorge; 'daar lê 'n kloof tussen hulle' – 'there's a gulf between them'. If you are at the bottom of the ravine or kloof, you may refer to the line of exposed rock at the top as the 'kloof'; or if you are near the bottom of a hill or mountain, and there is an exposed line of overhanging rock some way up the slope, providing shade and shelter, that could also be a 'kloof'; the exposed rock line at the top would in contrast be the 'krans'. 'Kloof' is also anglicised (to rhyme with 'hoof') as in the geographical name for the suburb just inland from Durban through which 'Kloof Gorge' runs.

Bult: knoll, hummock, hill(ock); ridge, rise, rising ground; 'dis net oor die bult' – 'it's a stone's throw away'; 'ons is oor die bult' – 'we are over the worst/we have weathered the storm'. In my experience, it is usually used to refer to a flattish ridge, which can be quite substantial in height but is not very steep in ascent.

Spruit: tributary, side-stream, creek, influent, brook, watercourse, feeder stream, spruit; frequently used for small streams that may not flow throughout the year. It is interesting that a river, which may typically go from being a spring, a trickle, a streamlet, a headwater, a stream, a river, and finally an estuary, generally only has one descriptor. The Lion's River, despite its grandiose name, is pretty small, more stream than river. In contrast, the Riflespruit is one of the largest trout rivers in the country, on a par with the Umzimkulu River, despite being labelled a 'spruit'. Fly fishers will differentiate sections of rivers into headwaters, freestone streams, canal waters, but even this leaves a lot to the imagination. A spruit is not a river or stream or rivulet. It is – *mos* – a spruit.

Rand: ledge, brink, lip, edge; yet in the name of the place 'Die Rand', it means 'The Reef', the location of South Africa's original gold mining industry, and in this respect it gives its name to the South African currency.

Berg: mountain; mount; used in a generic sense for all high, craggy geological protrusions; as in English, it has extensive figurative associations. It is used informally and in English with upper-case B to refer to the Drakensberg ('the Berg', rhyming with 'burg'). More specifically, it would refer to the Drakensberg range in KwaZulu-Natal and possibly Lesotho. The Drakensberg range extends much further south than KwaZulu-Natal, but I have never heard people from, say, Somerset East, referring to 'the Berg'; rather for them it is 'the Drakensberg'. While the dictionary definition refers to 'mountain' in the singular, in place names – Winterberg,

Cederberg, Magaliesberg – the term means 'mountains' or more accurately 'mountain range'.

Trek: besides its meaning to 'pull', which is obviously associated with 'travel' in animal-drawn transport, it means 'journey, travel, go, march, trek, migrate'. 'The Great Trek' refers to the historical event with profound political associations and implications in South African history. Idiomatically, in English and Afrikaans, it is used for a major, perhaps even life-changing, journey – a journey across 'wild' terrain; or flippantly to a journey that one does not wish to undertake and which is going to be demanding ('a real trek').

Krans: cliff; (krans) krantz; precipice; rockface; crag; high rock; 'kransberg' means 'escarpment mountain'. It was a powerful symbol of white (Afrikaans?) South Africanness in the apartheid national anthem 'Die Stem'; but the line 'die kranse antwoord gee' has since been included in the new national anthem of post-apartheid South Africa.

Veld: veld; field; grazing; pasture; vegetation; country; hunting ground. It is interesting that the emphasis in the dictionary definition is on the 'agricultural' over the 'wild', which is emphasised so often in literature (hunting coming last in the list). It may also suggest a place away from the city/town, or distinguish uncultivated from cultivated land.

Bosveld: bush country; bushveld; bushland. Used especially of dry terrain with low, dense and generally thorny vegetation. The English derivation 'bushveld' can also include more lush landscapes, such as those in the north of KwaZulu-Natal, known as coastal bushveld.

Fynbos: scrub, shrub, fynbos, especially of the *Euryops* genus. It is apparently of Dutch origin referring to the fine (fyn)/small leaves

73

and flowers of the Western Cape coastal vegetation; as much in usage in English as in Afrikaans.

Hoek: simply means 'corner' or 'turning'; or 'narrow glen'. It appears in numerous place names – Keiskammahoek; Bulhoek – in which it refers to a sheltered, flattish place suitable for human habitation (on a small scale) amid mountainous terrain. I cannot think of an analogous English term.

Nek: literally, 'neck'; but by association mountain pass or saddle (as in Brook's Nek; Naude's Nek). Possibly originates from pre-pass days, when the 'nek' was the only route across a mountain range. More than just a 'pass': a steep road snaking through a mountain range. To say that you had driven to the town of Rhodes via 'Naude's Nek' in the rain is to evoke in the mind of the listener a sense of heart-in-mouth trepidation. 'Naude's Pass' does not do the same. And then, just when you think you have a handle on these terms, human geography throws you a curveball like 'Kloofnek Road'.

Vlei: hollow, marsh, swamp, bog, quagmire, slough, moor, small lake. Most frequently used for something referred to as a wetland, or to a shallow body of water with vegetation surrounding it or in it, the kinds of marginal bodies of water that abound in this water-scarce country. Would not be used to refer to a body of water of substantial depth. Lake and dam are too expansive; marsh does not suggest the free-standing water in a vlei.

Vlakte: plain or flat stretch of land. A defining feature of the South African landscape in the old national anthem, 'Die Stem'. A vlakte would be a good place to cultivate crops, or to allow livestock to feed safely. Associated with expanse, but also in poetry with desolation.

Windstil: calm or windless. Wonderfully oxymoronic expression, which suggests both wind and stillness, so that one appreciates better the relief from the wind.

Fontein: fountain; spring. Also used in the same way as 'oog' or 'bron' to refer to the source of a river. A fontein can be an important water source, rather like a well, hence its frequent appearance in place names, as it provided the reason for settling there (Rietfontein, Bloemfontein, etc.).

Wild: game, wildlife, venison, chase, quarry. The term has a similar breadth and emphasis on function as the African term *nyama*, discussed in the next section. Interestingly, as an adjective applied to humans, the meaning is generally negative.

Koppie: hillock; knoll. But more than just this. Frequently a landmark: 'Just past the koppie'; or 'Koppie Alleen', the only elevated piece of ground anywhere near the Free State town of Welkom.

Suurveld/Soetveld: sweet veld/sour veld. Terms identifying specific types of grazing, suitable for different animals under different conditions, or requiring particular farming techniques.

And there are many more: *rooigras*; *steilte*; *laagte*; *gat*; *brak*; *drif*; and so on.

* * *

We walk hand-in-hand with the weight of words between us. Sometimes the sea provides succour. Today the waves break with grim ferocity, wind scours the sand, and rotting kelp on the beach tells of wild storms at sea.

* * *

I want to turn now to African languages. Writing about representations of landscape in the southern African context, Elsie Cloete has explored the contrasting possibilities and limitations of African and European languages (primarily English) in understandings of environment, both historically and in the present. She offers keen observations on widely differing assumptions about human-environmental relations in different language systems and their political and economic implications. And she points out in this regard just how poor English is as a language to engage with African landscapes. Though she does not use the term, her work has considerable implications for 'rewilding' language.

Cloete points out that many African languages use the term *nyama* (meat), or a variation of this term, to refer to a wide range of wild animals: for example, *mnyama* (Kiswahili); *inyama* (isiZulu; isiXhosa; Xitsonga); *nama* (tshiVenda; siPedi)) or *nyamatsane* (Sesotho/ Setswana).[27] This is not to suggest that animals were not identified by species – they were – but that they were also generically lumped together as food resource. The tendency to prize use over taxonomy is something we will encounter a little later in descriptions of landscapes. Cloete points to the irony that the word *nyama* 'became locked in the printed medium of indigenous-language dictionaries in the nineteenth and early-twentieth centuries at almost exactly the same time that colonial authorities declared that wildlife as a source of meat for indigenous Africans was illegal'.[28] While 'game meat' had previously been a food resource for African and Khoisan/Khoikhoi peoples, it became legislatively off limits. As Cloete points out, even now:

A man who is caught trapping an antelope to feed his starving family will be charged under current law. In his defence, he will have no legal recourse to a counter-memory or meaning-system from the past and which is available in his own language. In the twenty-first century, *nyama* represents an unacknowledged 'extinction of experience' and will not be legally viewed as an intrinsic part of a cultural history.[29]

One could compare the experience of the nineteenth-century /Xam Bushman //Kabbo, who was part of the group of prisoners whom Wilhelm Bleek and Lucy Lloyd interviewed while they were being held at the Breakwater Prison in Cape Town. His crime was apparently cattle theft, an act that had become punishable in an economic and legislative regime in which animals were personal possessions rather than a communal resource.

Cloete points out further in this regard that the key terms 'wilderness' and 'bush' 'have no meaning-system equivalence [. . .] in most of southern and eastern Africa's indigenous Bantu languages. This, should misunderstandings arise, can have significant implications in terms of conservation, social justice, and indigenous knowledge systems.'[30] In this respect, describing a landscape as 'bush' is probably the equivalent of reducing the complex range of raptor species to the catch-all 'hawk'.

Cloete then examines some of the origins of notions of 'wilderness'. In Germanic languages, it referred to 'the deep, dark, and dread-full forests of Northern Europe', in which people did not reside, and which were the places of myth and legend. Southern Europe was more densely populated, and the Romance languages do not have the equivalent term. She quotes Roderick Nash on the fact that the closest in Spanish is the term *falta de cultura* (lack of cultivation), whereas Italian uses *scene di disordion o confusion* (place of disorder or confusion).[31] With the rise of industrialisation, the Romantic movement especially 'came to see the wilderness not only as forest land but also as those remaining places unspoilt by factories, slums and human degradation'.[32]

Within African languages the closest equivalents to the term 'wilderness' are: *indle* (isiZulu: 'the space just outside the hut or kraal where one goes to relieve oneself at night'); *porini lisilolimwa wala kuishi watu* (Kiswahili: 'a grass plain where people do not farm (but could pasture cattle)'); or *muzyonde* (Leya: 'a place that is not used for cultivation but where mostly animals (including livestock) may be found').[33] 'Wild Africa', Cloete says, translates in Setswana and Sesotho to *Afrika e hlala* ('home for the natives of Africa').[34]

The notion of 'nature' or 'wilderness' 'out there' is not evident in such African languages and cosmologies, and yet it is difficult – using the medium of English, for example – to avoid dualist distinctions between 'society' and 'nature': the language that has always seemed to me so supple and capable of the finest distinction and texture suddenly feels clumsy and inadequate in this respect.

Let me illustrate. The book *Mammals of Southern Africa and their Tracks and Signs* by Lee Gutteridge and Louis Liebenberg is a fine source for anyone wishing to learn how to read the complex signs and tracks of animals. The detail, nuance and registering of seemingly minute differentiators in the sketches, photographs and descriptors are extraordinary. They identify seven major biomes in southern Africa, defined by their major vegetation types – Forest; Fynbos; Grassland; Nama Karoo; Savanna; Succulent Karoo; and Thicket[35] – and then proceed to differentiate habitats. They identify nineteen for the entire region, some identified by characteristics such as 'Wetland and Swamp' or 'Bushveld', most by geographical name, such as 'Drakensberg Mountain and Lesotho Highlands', 'Limpopo Valley', and some that seem so generic that one wonders about their usefulness, as in 'Forest', 'Large Rivers' or 'Coastline'.[36]

In stark contrast, drawing examples from the Shangaan peoples in eastern Zimbabwe, whose geographical area would probably span the equivalent of one of Gutteridge and Liebenberg's more narrowly demarcated 'habitats', Cloete lists examples of words used to differentiate a wide range of features of landscape within the region.[37] They reflect more intimate dwelling perspectives:

Kutluma:	thicket – favoured as hiding places by hyenas, leopards and lions.
Mabhiripirini:	riverbank gulleys – where pythons are found.
Mabvungurhi:	canopied areas – where owls rest during the day.
Ndovolo:	area of black fertile soil.
Mathlivi:	stone depressions which collect water in the rainy season. Hunters, baboons and birds drink from them.
Thlaveni:	area of red, sandy soil – it is hard to bicycle because of the sand. Good for growing groundnuts and bambara nuts.

Magumbitsini:	thicket that is hard to penetrate.
Thangava:	portion of the field in which women grow their own crops (such as groundnuts, beans, sweet reeds and water melons).
Tipala:	salt pans – where evaporation leaves salt crystals. The salt is often gathered and sold; also: plains – favoured by impala during moonlight as it is hard for predators to attack them.
Marhimakule:	outfields – where people dig for tubers (*phomwe*) in drought years.
Chawunga: (*Chavunga?*)/ *mananga*	remote, quiet and fearful area where only birds and wild animals are found – popular with foreign visitors.[38]

Cloete points out that '[p]eople living in the so-called bush have developed a lexicon of terms that is based on experience and a careful reading of the environment. An inexperienced human blundering into a *kutluma* (a thicket where predators hide) could end up fatally misreading that environment!'[39] As suggested earlier, this is a lexicon concerned with use rather than taxonomy – knowing that thickets harbour predators might seem more significant in this context than being able to name each plant species that comprises them. Cloete does, however, note with sadness that many of these terms appear to be little known by people now, as their concept of landscape has been flattened instead by English into a generic concept of 'bush', particularly at schools.

The shift towards English that Cloete describes is not the aggressive, bellicose process articulated by Sir Harry Smith earlier in this chapter, but the effects are no less pernicious: 'Talk about the natural environment in particular remains westernized and monolingual, with little cognizance of the ranges of knowledges and practices embedded in the various African languages.'[40] At its worst, the abandonment of indigenous languages for English can lead to people inhabiting an environment without an adequate language or knowledge system with which to apprehend it; they become estranged within 'their own' world:

In rural areas where indigenous languages are mostly 'place-specific', such a 'language shift' to English at school puts at risk traditional ecological knowledge where the prism of a single system of Western knowledge (represented by English) displaces knowledge-systems outside the physical boundaries of the school. Ultimately, this creates the bizarre situation that a person indigenous to a place becomes a kind of 'foreigner' in his or her own country.[41]

Language has not just been 'dewilded' in this process; people have been 'deworlded'.

If we are serious about 'wilding', we need to shake up the languages we use to talk about the worlds we inhabit. It is a process of learning or relearning, a potentially invigorating antidote to the anodyne language that characterises so much global discourse about humans and our environments, and which is so far removed from the blood and urine, the pain and rage, the piercing beauty and wonder of life on this miraculous planet we call home.

* * *

Father used to tell me that, when lying in wait for a porcupine, at the time at which the Milky Way turns back, I should know that it is the time at which the porcupine returns. Father taught me about the stars; that I should do thus when lying in wait at the porcupine's hole, I must watch the stars; the place where the stars fall, it is the one place which I thoroughly must watch. For this place it really is which the porcupine is at, where the stars fall.

I must also be feeling (trying) the wind. Things which I should watch, father in this manner taught me about, things which I should watch. Father said to me about it, that I should not watch the wind (i.e. to windward), for the porcupine is not a thing which will return coming right out of the wind. For, it is used to return crossing the wind in a slanting direction, because it wants to smell. Therefore, it goes across the wind in a slanting direction, because it wants to smell; for its nostrils are those which tell it about it, whether harm is at this place.

Father used to tell me, that I must not breathe strongly when lying in wait for a porcupine; for, a thing which does not a little hear, it is. I should also not rustle strongly; for, a porcupine is a thing which does not a little hear. Therefore, we are used gently to turn ourselves when sitting; because we fear that had we done so (noisily), as it came, it would have heard.

//Kabbo. 'Habits of the Bat and Porcupine'.[42]

5

Wild Seams and Margins

And she walks along the edge of where the ocean meets the
   land
Just like she's walking on a wire in the circus.
                    — Counting Crows, 'Round Here'

Wildness is not something 'out there', something beyond the urban humdrum that defines the lives of most of us. We have wildness within us, and we also constantly approach or cross seams or margins of the wild, even in the most apparently urbanised spaces. In this chapter I look at the ways in which we move, often even in our daily lives, across the zones of what we would conventionally call wild and urban or wild and tame.

There are probably few beachfronts as domesticated as Muizenberg. Surfer's Corner is home to coffee shops, surf shops, surf schools, an ablution block, water slides, and a car park in which yuppie SUVs and the rusted vehicles of the old longboarders vie for space. It is a bustling, lively, urban terrain.

But zip up your wetsuit, grab your board, walk across the beach into the shallows and you cross an important line. As you paddle out through the crowds of surf school trainees and newbie surfers, you are entering another world, with different rules. You would probably have checked the Shark Spotters' flag before entering the water, to see what spotting conditions are like, and glanced up to the station on Boyes

Drive, from which they survey the seascape, ready to sound the alarm when a shark approaches the surf zone. Approaching the backline, you can look back to the shore and see the building lines of Surfer's Corner, and the compressed domesticity of Muizenberg shuffling up the mountain slope. But your eyes may also flit more anxiously to the waters around you and to the depths beneath you, because here you may still be Carlo the accountant or Sandra the academic, a surfer of better than average ability, but in a real sense you are also a very frail mammal bobbing about on the surface like the poppers used by fishermen to attract game fish – potentially prey to the great whites, which frequently visit this beach in search of food. You may retain something of your daily identity in interactions with other surfers at the backline, but you may also form friendships around surfing that you could not sustain out of the water. That is, if you're a good surfer, lucky, or a local, because the competition for waves can be decidedly feral.

Much of the thrill of surfing, of the heart in the mouth drop, the speed along the open face of the wave, the hanging on to see if you will make the section, is an encounter with wildness. You face the possibility of drowning in big surf, of a decidedly one-sided encounter with an apex predator, or of injury in a bad wipeout, but still you are drawn inexorably across the car park and into the ocean.

In his book *Alone: The Search for Brett Archibald*, Brett Archibald narrates his experience of remaining afloat for 28-and-a-half hours in the ocean off Indonesia, after having fallen overboard from the yacht that was taking him and some friends on a surf trip. As surfers, he and his friends share a 'relationship with the ocean [which] is one of careful balance: it can be entirely captivating, but it also demands respect'.[1] Yet Archibald's relationship with the ocean faces severe tests during his ordeal. Miraculously he survived, and he captures the sense of being catapulted from the relative sanctity (they were all suffering from food poisoning) and camaraderie of the yacht into the vastness and unpredictability of the ocean: 'I imagine the cavernous depths beneath me; the dreadful secrets they hold. The ocean, while I've loved it since I was a child, is still unknown. Unknowable. It's alive with creatures that I might well encounter soon. I've never felt this vulnerable before. Ever.'[2] A little later in the narrative, he comments: 'I am in the deepest part of the ocean, unlit, enigmatic, with hundreds of metres of briny deep stretching beneath me [. . .] This is another world, a place I don't belong.'[3]

* * *

Campus has been closed for more than a month due to violent student protests, when a message from the university executive says that a temporary lull means that staff can access their offices if need be. I drive into campus amid tight security. The verges are overgrown, leaves, broken branches and litter are strewn around, and the usually bustling roads and walkways are eerily empty. Police and security vehicles prowl slowly around, watched impassively by small knots of students. Occasionally, a lone student scurries by, head down. There is an overwhelming silence, broken only by the rasping of the crows banking and circling overhead, the incongruous cooing of doves, and the nagging and tugging of the wind. Getting into my building past a line of security fencing and somnolent guards proves impossible. Broken windows and scorched walls tell of the recent violence, and my nerves are jangling as I seek some point of entry. The place feels feral.

Eventually, security allows me through a glass door. I walk in semi-darkness through echoing hallways. Loose windows clank and rattle in the wind. There is not another soul in this huge expanse of lecture theatres, offices and seminar rooms, and the place feels abandoned. Unlocking a security gate and groping my way down a dark passage, I enter the staff toilets. There is an explosion of flapping and wing beats, and a dove erupts from one of the cubicles, launching herself at the window above the basin. I manage to grasp her, and tilt the window open enough to push her through the opening. Covered in dove shit, I then start to take in my surroundings. There are twigs and feathers everywhere. Clearly she has been building a nest, it seems of mammoth proportions. Dove shit covers just about every surface, and the urinal is filled with more twigs and branches. In the space of a few short weeks, she has begun rewilding this place.

*   *   *

Liminal. From the Latin *limen*, meaning 'threshold'. The piers that thread Durban's beachfront are liminal spaces. Built ostensibly to 'control' the sea – to put a halt to strong rip currents that scour away sand and swimmers – they are now partially rewilded. The bases of the pillars on which they rest have been colonised by mussels, crabs, barnacles, crayfish and seaweed. Woe betide any surfer or swimmer who gets washed under there.

The piers have created the left and right peeling breaks beloved by surfers. Making your way out along the concrete walkways is the closest non-surfers will get to the experience of being out at the backline. Should you avoid the attention of the authorities and walk out to the end of the pier with your board to save yourself a paddle out, you can launch yourself directly into the surf. You are transported in a split second of freefalling from the world of ice creams and cellphones into a surging mass of water. Fitness, skill and wits are your only allies, and your strength seems puny.

Today, there is a gentle north-easterly blowing, and a shoal of shad has drawn the fishermen. The end of the pier is a forest of rods, with the shad too long a cast from the beach, and only the reach of the pier into the sea providing access. There is controlled chaos, with the fish grabbing baits as they hit the water, and feverish rebaiting between casts. Practised lifts bring the fish out of the sea directly into fishing bags, out of sight of any Parks Board officials who might be policing minimum size and catch limits. An hour later it is over. The shoal has moved off, and there is not a fish or angler to be seen. Only sardine heads, entrails and the odd scale hint at the encounter.

*   *   *

This is sea that can kill you quickly, winter or summer. Beyond us, strung along the coast to the north, are all the fishing towns, Peterhead and Fraserburgh, Buckie, Portsoy and Cullen, places that know well the closeness of sea and death. — *Esther Woolfson,* Field Notes from a Hidden City

* * *

We still depend to a significant extent on negotiating the wild/ domesticated interface for sustenance. At a more subtle level, which I discuss in a later chapter, growing vegetables or farming animals involves harnessing wild, biological processes. In a more obvious way, catching fish at sea involves an often precarious, dangerous venture into a hostile environment. In his well-known poem 'Ken Jy die See?' (Do You Know the Sea?), Uys Krige imagines the response of a fisherman to a customer who complains that the fish he is selling is too expensive. His account draws a stark contrast between the protected domesticity of the purchaser, and the treacherous and hazardous lives of the fishermen:

Ken jy die see, Meneer, ken jy die see?	Sir, do you know the sea?
Hy lyk nou soos jou voorstoep blink geskuur En kalm soos min dinge hier benee, Maar hy's gevaarliker as vlam of vuur. Dan sê jy nog, Meneer, die vis is duur . . .	Now it looks like your front shiny porch, And flat-calm, like so few things here below, But it is more dangerous than flame or fire. And yet, Sir, you still say the fish costs too much . . .[4]

The fisherman gives a stark account of the perils of his life:

Was jy al van jou bootjie soos 'n veer gevee	Have you ever been swept off your boat like a feather
Deur 'n grys golf hoog soos 'n tronk se muur?	By a grey wave as high as a prison wall?
Wat help dit om te spartel en te skree 'Nee! Nee!'	What does it help to struggle, and to scream 'No! No!'
Sluk jy eers daardie waters sout en suur?	You just swallow salty, bitter water?
Dan sê jy nog, Meneer, die vis is duur . . .	And yet, Sir, you say the fish costs too much . . .

The poem ends with the image of a bereft mother, always looking anxiously out to sea for the wind to bring back her lost sons:

Sien jy die krom ou vroutjie daar, Mevrou Mathee,	Do you see that old lady over there, bent with age, Mrs Mathee,
Wat telkens ver, ver oor die golwe tuur?	Always gazing far out across the waves?
Sy dink die briesie bring Haar seuns betyds terug vir tee.	She thinks the wind will bring Her sons home in time for tea.
Hul slaap al drie agter die kerkhof muur.	All three lie behind the church yard wall.
Dan sê jy nog, Meneer, die vis is duur . . .	And yet, Sir, you say the fish costs too much . . .

Douglas Livingstone's poem 'Sonatina of Peter Govender, Beached' also explores the perils fishermen face, but his interest appears to be more in the psyche of a life of fishing and the identities that form around it.[5] For this old Indian fisherman, fishing is a compulsion ('I had to fish: / first, surf; then the blue water marlin') and a hard-earned, well-practised skill (Snyder would approve): 'My prime as oarsman: / heroics of the offshore boat, / catching all that steel slabs of sea could express.' Fishing and the sea are a source of ambiguous and contradictory knowledge:

Things learnt from the sea
– gaffing the landlord, the week's debt,
scooping in the crazed white shads,
twisting the great transparent mountains
past a wood blade – ?

But finally, for the 'curt' old man who has only 'monosyllables for strangers' and whose 'porpoise-wife has gone', the daily encounter with the ocean encapsulates the paradox of living with the awareness of the potential imminence of death, which haunts us all: 'Contempt for death is the hard-won / ultimate, the only freedom . . . . / not one of the men I knew could float.'

* * *

Bodyboarding in the Eastern Cape, I have had many encounters with dolphins. There is the heart-stopping moment when you first see a fin, and then an overwhelming sense of relief as it dips and rises in the unmistakable cadence of dolphins, and you start to spot the fins of the others in the school. (Yes, the shape of dolphin fins differs from those of sharks, but I would defy all but the most assiduous marine biologist to register that in the panicked millisecond in which you see the fin for the first time.)

Often they just swim on by, and we slide off our boards and listen for their sonar squeaking with heads under the water. But on one afternoon forever etched in memory, they decided to keep us company. A friend was out on a windsurfer, and the dolphins chased the board, taking turns to jump across the nose. A very large individual swam in to have a good look at me, right up close. While I knew rationally that I had nothing to fear from it, I felt decidedly frail and humble as this massive creature rolled past me, with a girth I would have battled to encompass with both arms, and a presence that was huge and undeniable. It swam lazily away from me, as I sat on my board, then turned and swam straight back at me at high speed. At the last moment it ducked underneath me, and resurfaced a few metres away. Try as I could, I could not see it at all as it passed beneath me. I was shaken and exhilarated.

They appeared to tire of our company, and not knowing what I do now, that sharks often follow schools of dolphins to pick off the young or to pick up the scraps they leave behind when feeding, we carried on surfing. As I paddled into a wave, two sleek silver-grey shapes emerged through the face of the wave about a metre from me – two dolphins surfing this wave with me. In the lore of surfers, it was my wave, as I was closest to the break, but I didn't think the dolphins would have much regard for human claims to right of way in the ocean. Besides, these beautiful creatures dwarfed me in size. I gave them best, pulling out of the wave, and was rewarded with the sight of one of them flicking itself into the air as the wave finally closed out.

* * *

The human creation or negotiation of liminal spaces can at times be clearly evident in physical, visible terms. In contrast to many urban gardeners who cultivate vegetation right up to their front doors (though as I point out in a later chapter urban gardens also involve a process of mediating the wild), in almost all rural African communities, there is a cleared, regularly swept area of bare earth around each dwelling, so that unwanted intruders such as snakes or rats can be clearly seen approaching and quickly despatched. The area of bare earth marks the transition from indoors to outdoors, from domestic space to 'what else is out there' (which is a working definition of 'wildness' for many).

But the lines that we draw between the wild and the tame, or the wild and the protected, may frequently be notional, though their consequences may nevertheless be momentous. Let me take two contrasting examples.

On a school trip in the early 1980s to the Okavango Swamps and surrounding game parks in Botswana, we arrived at the end of a long, gruelling drive at the gates of Moremi Game Reserve. I mean 'gates' in the literal sense as there were two gate posts and a closed gate across the dirt road, but no fence on either side of each gate post. Even a suburban sedan could have driven through the veld on either side of the gate to access the reserve, let alone the four-by-four vehicles in which we were travelling. It was as notional a 'gate' as I have seen. For the law-abiding school teachers who were leading the trip, it was an absolute barrier, and they refused to 'enter' the 'game reserve' without the gate being unlocked. As we could not get to the campsite at which we had planned to overnight, Plan B was then to set up camp 'outside' the reserve. That night around the fire, the air shuddered with the roaring cough of a male lion of uncertain distance away. As we were apparently 'outside' the reserve, and there was no sign of rain, many of my classmates opted to eschew their tents and sleep under the stars, to no apparent concern from the teachers. Being a little more schooled in 'the ways of the wild' from trips to game farms and the like, and not imagining for a second that the animals would respect the imagined boundary in the way our teachers apparently did, a friend and I took the wiser option of sleeping on the back of the large truck that was carrying our baggage. We awoke in the morning to the rude shock of a pack of hyenas having raided the camp, dragging off unwashed plates, cameras and camping gear, which had been centimetres away from the heads of those sleeping in the open. To this day I do not know how no one escaped injury.

In the second example, the line is also notional, but in a profound sense 'real'. The film *The Great Dance*, produced in 2000 by Craig and Damon Foster, explores the hunting practices and belief systems of the !Kung people of the northern Cape and southern Botswana in the 1990s. In a truly compelling sequence in the film, three men engage in one of the most physically and spiritually demanding of hunts, in which they run a kudu to death. The concept of a human being able to run an animal to the point of exhaustion is probably beyond the imaginative grasp of most readers, especially city dwellers, but that is exactly what is done. As the physical pursuit of the animal unfolds, a process of spiritual identification and transposition takes place, as the hunter 'becomes' the animal, and wills its exhaustion even as he experiences his own. At the end, when the animal is speared, at the level of being it is not clear who kills whom.

The hunt by running is one of the most extraordinary feats to witness, but there is an unusual turn in the first hunt recorded in the film. As the hunt begins to reach a pitch of physical and spiritual intensity, the animal crosses an apparently innocuous clearing between some bushes, and the hunters stop dead in their tracks. The bush into which it disappears looks no different from anything that has been seen on the hunt so far, but the animal has crossed into a 'reserve', in which it is no longer 'prey' but a 'protected animal' (it is an area in which fences have been dropped so as to allow animals to migrate freely). Presumably rather than risk substantial punishment, the hunters have no choice but to abandon the hunt at this advanced stage. A few metres of unremarkable veld fundamentally change human-animal relations in terms of identity, use and value.

At other times, the boundaries are more clearly demarcated, though no less troubling or contradictory in their implications. In 2016, there was a great deal of debate in the media about an incident in which a four-year-old American boy visiting a zoo with his mother fell into the gorilla enclosure. The gorilla picked up the child, and the zookeepers, fearing for the boy's safety, shot the animal. Much has been said about the negligence or otherwise of the mother, and I have no wish to enter that debate. What is intriguing for me, instead, is the way in which arguments for or against the gorilla's being killed were framed: in terms of the fundamental 'line' that was crossed when the boy went over the railings, of his crossing over into a radically different, 'wild' world.

I was on a research fellowship at the Stellenbosch Institute for Advanced Study at the time, working on this book, and the opinions of those at the lunch table – all leading figures in their fields of thought – on the topic were intriguing. While information suggests that the gorilla in question had been born in captivity (and was 'named' – Harambe – more of which in a later chapter), a central and unquestioned assumption for many debating the issue, on both sides, was that the gorilla was a 'wild animal'. Those who supported the animal's being killed framed the argument as such: the gorilla was a wild animal that was a danger to the child, and the protection of the child was paramount, so the animal had to be shot. Its 'wild nature' was dangerous. In this argument, human rights take precedence over those of animals. Those who felt it should not have been shot argued that the gorilla was a wild animal acting 'as it should', so it should have been spared. This second position is an argument about culpability, and the animal is granted rights comparable to those of humans: it is the humans who are culpable, not the gorilla, so it does not deserve to die. But both arguments rest on the assumption of the animal's wildness.

In the heart-stopping moment in which the small boy dropped into the enclosure, he is seen to have crossed a line that is fundamental to our understanding of the contradictory and controversial nature of the urban institutions we call zoos: they comprise pockets of (relative) wildness in urban spaces. The degree of the qualifier 'relative' is crucial here. Leaving aside the extreme examples of zoos that incarcerate animals in cramped, unhygienic and foetid conditions, the most pressing ethical question asked in relation to zoos is whether it is 'right' to keep 'wild' animals in 'captivity'. Confining a dog to a suburban garden sufficient for its exercise needs, with proper nutrition, would not raise similar objections to keeping a hyena or jackal in an enclosure. It is also intriguing that the confinement of animals such as monkeys, apes and chimpanzees, whose behaviour and physiology more closely mirror our own and so may 'temper' their perceived wildness, seems to be less at issue in public perceptions of zoos than the incarceration of emblematically 'wild' predators like the big cats. It is the assumed wildness of the animals that is at issue here, and the need for security measures to 'protect' humans from the animals confirms their 'wildness' in the potential for harm. So in the understandings of many, the boy fell into a pit of wildness, in much the same way as Daniel was thrown into the lions' den in the biblical narrative, even though this occurred in downtown Cincinnati. Wild edges, wild seams . . .

Henrietta Rose-Innes captures the sense of a zoo or animal park as representing a wild 'heart' within an urban space in her novel *Green Lion*, set in Cape Town. The lion in the novel becomes a source of fascination, freedom, fear, sexuality, energy and transcendence for the protagonist Con, who works at a captive breeding facility in the city, and confronts many of his own personal problems through his relationship with a lioness called Sekhmet (an Egyptian warrior goddess), which had mauled its previous 'keeper', a friend of Con's. Even after his first encounter with the lioness, he begins to experience a transformation within himself akin to that described by George Monbiot when he hefted the dead deer onto his shoulders:

> Con felt strange. Sharper. The world had a fine grain, an artificial brightness to it. He knew what it was: the kick of big-predator adrenaline, working its way around his system. Maybe he would run all the way home; it wasn't too far, down past the old wildebeest enclosures and along the highway.[6]

When he does arrive home, having brought his mauled friend's work belongings with him in a rucksack, he experiences a sense of the feral:

> He sniffed his palm and caught a whiff of something wild. The reek of the lion cage was on his skin, or perhaps emanating from the rucksack. He regretted bringing the smell into the flat, into Elyse's intimate air. Like a dog bringing roadkill to the door, unsure if he'd be smacked or petted. He put the bag down in a corner, stripped off his sweaty clothes and put them straight in the washing machine. He needed to wash.[7]

I am not entirely convinced by or satisfied with the narrative that then unfolds in the novel, though it is still in my opinion well worth a read. The lion remains a brooding, elemental presence in the narrative, and its breathing, scent and roaring appear to inform the protagonist's very sense of who he is becoming.

A rather different scenario involving a human in a gorilla enclosure played itself out in the Johannesburg Zoo in July 1997. Fleeing from a crime scene adjacent to the zoo, Isaac Mofokeng sought to evade police pursuit by entering the zoo grounds and jumping into the enclosure containing two gorillas. The male, named Max, grabbed hold of Mofokeng, who shot him three times. Police then entered the enclosure and wounded and apprehended Mofokeng, though three of them suffered injuries at the hands of an understandably enraged Max. Following emergency surgery, Max returned to good health.

In a country bedevilled by crime, Max became something of a celebrity, with public opinion firmly on the side of the gorilla, and Mofokeng being adjudged to have gotten his just deserts – a kind of 'jungle justice' or 'wild law'. The Johannesburg Press Club named Max 'Newsmaker of the Year', and the zoo subsequently erected a statue of Max. A news report on Max's death several years after the shooting was entitled 'Crime-Fighting Gorilla Dies', and concluded that '[n]ational heroes come in many shapes and forms'.[8] Mofokeng was found guilty on ten counts, including rape, robbery and housebreaking, and received a 40-year jail term. Intriguingly, five of those were for 'malicious damage to property' for the shooting of Max, who – despite becoming something of a folk hero – was legally adjudged to be a piece of property, valued at R2.5 million. Mofokeng subsequently apologised for shooting Max, but not for the rape . . .[9]

*   *   *

With some it is water shrugging, bunched and oily
at the quayside – the cold welcome of lewd carpets;
for others, the pineal-sucked lure, dragging dizzy
and out from windy skyscraper parapets.

With him it was the tiger: beautifully slack;
indifferent; sleep and captivity thinned;
lying on a fat pole like a striped rug, back-
legs adangle, forepaws crossed under chin.

He even learnt a few words of Bengali (culled
from Tagore) and leapt the ditch to press
long and urgently at the bars, mad to scratch unpulled,
tortoise-shelled and round furry ears.

Angry keepers and others ordered him back and he
went, backwards, arms out, aching and bent
about air the size of a tiger, and thought of his granite-
faced and quite unfurry apartment.

To shed his love one night he broke in, sat his
city trousers on a foliage-encrusted stone wall,
jumped running for the beloved bars, fumbled latches
and reverently entered the shrine though the feed-door.

For perhaps one second he felt it, face buried in rank
cat's fur: the sleepy response. Then the rasped purr
meshed with metallic springs. The barrelling flanks
pumped an outraged blast from alien vaults of power.

They found him on the floor early next morning, his head
a split and viscid watermelon; loosely the wet tufts
of combed brains spilled, his smile quiet through the red;
beside him, for warmth, the cosy sprawl of his love.
                    (Douglas Livingstone, 'The Zoo Affair')

*   *   *

Seashores and coastlines are among the obvious seams along which we negotiate wildness. But some are much closer to home. Such as when two people remove each other's clothing to reveal bare skin.

But skin itself is a liminal zone. Our visceral response to the roadkill that we frequently encounter on our highways – the tangled mess of fur and blood, the inner organs that have burst through the covering of skin – may suggest our own latent (and frequently repressed) awareness that the outer wholeness that we perceive as the animal or human before us is a very small part of the story.

Skin is a large and wondrous organ that contains us, our organs, musculature, bodily fluids; it (mostly) conceals the processes and body parts that keep us alive. Beneath the sanctity of the skin the wild sustains us: processes function to minute exactness without our conscious cognition or control of them.[10] Skin creates the illusion of wholeness, of a body working in harmony with itself. But, as doctors, nurses or paramedics know only too well, the barrier between the composed self and the broken body is a few fragile millimetres deep. All too often they see precisely what lies beneath the skin. At other times, they seek points of entry to determine what is malfunctioning, through the skin physically with needle or scalpel, or virtually with X-ray, ultrasound or scan; or the orifices of the body provide access to scopes, probes, biopsies. And when we become ill, battles are fought within us, both aided and disabled by powerful drugs that can save at the same time as they wreak havoc with our biology. Like the electricity we depend upon but which can kill us, or the heat we harness for cooking but which can disfigure and maim, medicine tries to harness the wild in the name of sustaining life, or to curtail it when it threatens the body.

We all live with the sense that our own bodies may betray us; that we carry within us the seeds of our own destruction. It is an awareness captured vividly by the poet Douglas Livingstone, when he talked of the 'nocturnal terrorisms' of the 'crab' (cancer) and the 'clot' that threaten the frailty of the body, which is 'little but a glove / stretched from metatarsals to neocortex / on a stiffening frame'.[11]

As the desires, strengths, frailties and shortcomings of our bodies indicate, we live at a wild interface . . .

6

Wild Cities

Cape Town is that rare thing: a hot urban mess that has
not yet smothered the wild.
— Helen Moffett, 'A Tale of Two Cities'

Ask most people to represent the opposite of wildness pictorially
and they will generally give you an image of a skyscraper, city skyline
or suburban house. But cities are themselves also significant sites of
wildness, or perhaps more accurately contain within themselves
significant conditions of wildness, and many populations of wild plants
and animals find their homes in cities. In previous chapters I have
referred to birds of prey in cities, and one of the better-known examples
of this is the way in which peregrine falcons, which had been pushed
to the brink of extinction in most parts of the globe, have adapted
so well to urban environments, swopping cliff tops for skyscrapers,
ledges for balconies, and making good use of the abundant supply of
pigeons. At a colloquium I attended in 2010, an ornithologist from
the University of Cape Town pointed out that humans have in fact
significantly assisted the peregrines in their colonisation of urban spaces
in that they help protect nesting sites, and will frequently come to the
assistance of nestlings that fall to the ground, or birds that are injured.
When the two cooling towers near Pinelands in Cape Town were
imploded some years ago, the demolition date was set to accommodate
the breeding season of the pair of peregrines nesting there. The same

ornithologist pointed out that peregrines are so acclimatised to urban life, for example, that four minutes after the cooling towers fell, the two peregrines were copulating in their new nesting box nearby (which prompts the obvious question, 'Did the earth move for you too, dear?'). However, scientists have more recently raised some serious questions about the genetic health of certain urban raptor populations.

Other examples of 'wild' populations of animals in cities globally include large numbers of leopards in Mumbai (claimed by some to be the largest concentration of leopards in the world), a massive peregrine falcon population in New York (also claimed by some scientists to be the single largest congregation of these birds globally), vervet monkeys in Durban, hyenas in Ethiopian cities, large catfish in Rome (which apparently eat pigeons), massive flocks of kites in Kampala, porcupines in most African cities, and so on. In a recent book, *Darwin Comes to Town: How the Urban Jungle Drives Evolution*, Menno Schilthuizen has pointed to significant genetic and behavourial changes in urban dwelling species as they adapt to their new habitat, including, for example, crows in Sendai, Japan, dropping nuts in traffic so that the wheels of cars crack them for their consumption, and wingspan reduction in American cliff swallows that have colonised concrete road bridges.[1]

The categories of the 'wild' and the 'urban', the 'wild' and the 'tame', or the 'wild' and the 'civilised' – as absolutes – are particularly problematic when one considers cities, which are, of necessity, both. Though their presence may frequently be beneficial (such as owls controlling rodent populations), 'wild' animals in urban areas often appear in the press or in public debate as 'problems': raccoons in the US raiding litter bins; foxes in the UK doing the same, or killing pets; or monkeys or baboons entering homes in southern Africa. A recent report entitled 'Boston is Covered in Goose Poop and People are as Mad as Hell' lamented the problems associated with the rapidly expanding population of non-migratory Canadian geese in the city of Boston, which is presented as slowly submerging under goose shit.[2] The article points out that a single goose can consume 'up to four pounds of grass per day and produce as much as three pounds of fecal matter every day'. One is reminded that, but for the invention of the internal combustion engine, which replaced animal drawn transport, most major cities would long since have been knee-deep in horse shit. Intriguingly, many Boston residents report conflicting feelings of frustration at the pollution caused by the geese and concern that this frustration may lead to harm to the geese at the hands of the authorities.

The case of the baboons in the Western Cape is an especially interesting one, which shows up the inadequacies of legislation and the concepts on which it is based (the wild versus the tame or the human versus the animal) in dealing with environmental issues. There was recently contestation between the City of Cape Town and CapeNature as to who was responsible for the problem of baboons in residential areas. As it is a complex and intractable problem, both parties would understandably have wanted the problem allocated to the other side. Intriguingly, and also bafflingly, legal judgment decided that the baboons were a 'nuisance' (as in 'Commit no nuisance'), and so were the responsibility of the City, not conservation bodies.[3]

* * *

The garden of our block of flats in Durban comprises mainly patchy lawn and a couple of trees that offer little by way of food for the birds, so we generally see nothing but Indian mynahs and the occasional dove. A frantic to-and-fro scurrying from the study one morning suggests that our cat is up to something. Closer inspection shows that she has caught a pygmy kingfisher, exquisitely coloured with iridescent blue, red and washed mauve. We have never seen one in the wild. What it is doing in suburban Glenwood, miles from any body of water, we cannot fathom. And how our portly feline caught it is equally baffling. But it soon recovers, and takes flight from the bedroom window.

* * *

Esther Woolfson's *Field Notes from a Hidden City: An Urban Nature Diary* is, as its title suggests, a year-long account of wildlife in the city, in this case Aberdeen. It is a carefully researched, closely observed narrative that explores the multiple ecosystems in and around the author's home in that northern Scottish city, involving rats, birds, foxes, slugs, spiders, plants, and much else.

Something of the reorientation in thinking that she asks us to make is suggested right at the beginning, when she narrates the experience of picking up a fledgling pigeon from a snowdrift, which – as per habit – she takes home to raise before releasing it. (Though she takes the time in her narrative to distinguish between fledglings that have been abandoned, and so should be rescued; and those that have landed on their first attempt to fly, will be cared for by their parents where they are, and so should be left alone.) At base, she asks us the rather unsettling question, 'Are city pigeons (and humans) wild?'

> As I fed and took care of the bird, I began to think more about the expression 'return to the wild', the words used to describe setting a creature free to return to its own environment, to live amongst its own, a natural life. The wild in this case would most probably be the large sandstone church at the end of the lane, a neo-Gothic building of step-corbelling and intricate stone finials, around whose tall central spire crows and magpies engage in glorious aerial combat on windy days. It would be the roofs and gardens of this district, the handsome Victorian villas, long-established trees and wide, quiet streets only a short walk from the centre of the city. Hardly wild.

But, she continues:

> For all that, this was a wild bird – a wild, city bird although the words 'wild' and 'city' seemed difficult to reconcile. I began to think about wildness in relation to creatures who live in cities, about whether or not we consider them less wild than creatures

living elsewhere, or think of them as somehow a lesser part of nature itself. I wondered if the same might apply to humans, as if merely by being in a city, not only might our lungs be polluted but ourselves, our minds (and if we have them, souls), as if urban dwellers must by definition be over-avid consumers of the unnecessary, weakened by purchase, alienated in every way, distanced from a lost, admonitory Eden. Are we, I wondered, living lives remote from all that is natural, beneficial, wild, or are we as much part of the natural ordering of the universe as the wildest of things, moved by the same forces, as wild as anything else on earth?[4]

She points out correctly that city pigeons (or sparrows or mynahs or starlings, depending on the context) 'may be the only contact many urban people have with the natural world but our relationship with them seems changed by proximity, diminished by the very fact of their being here among us';[5] a sense that our own presence must immediately preclude their being 'wild', or more sadly perhaps a case of 'familiarity breeding contempt'. Her musing on the pigeon leads her to the possibility that it is a 'wild bird but an ordinary one'. She goes on: 'I looked up the definition of "ordinary" – *With no special or distinctive features, common, of ordinary rank, undistinguished, commonplace.* Which of us is any more or less?'[6]

As I have argued previously, an important part of understanding wildness, especially urban wildness, is knowing how to look and hear. What you may at first think is a pigeon turns out to be a falcon, because you look twice. That lump of lichenous wood where there was not one yesterday is an eagle owl. Or the frantic cheeping as you walk near a bush reveals a nest of hungry fiscal shrike chicks. It is a matter of reading the signs differently, perhaps more attentively, more fully. And when we do, we realise that '[o]ur streets, buildings, houses are shared, our gardens and our trees. In one city, there are more cities than we know, hidden cities inhabited by those with whom we share everything we rely on: food and light and air. In differing degrees, we share our vulnerability to the elements that shape and dominate our lives: cold or heat, wind and rain.'[7]

'There are more cities than we know . . .' We can talk of 'rat Cape Town', 'sea gull Cape Town', 'baboon Cape Town' or 'cat Cape Town', the latter recorded in a research project from the University of Cape Town that attached GPS recorders to domestic cats in order to track their nocturnal activities, and something to which I shall return later in this chapter. I will discuss rats and their behaviour in a separate chapter when I consider animals and ethics, but Woolfson points out that '[i]n the early 1980s, the French graffiti artist Blek le Rat drew rats all over Paris, to honour the rat as "the only wild thing living in a city"' (although that assumption is somewhat mistaken), and that in London 'Banksy portrays the rat as an urban character and hero, or anti-hero, as anarchist or gangster'.[8] Welcome or not, rats have been associated with human movement for much of our history as 'wild', unwanted, and hidden companions, and have also had a causal effect in our relationships with other species such as cats, which we have 'used' to control their numbers:

Both *Rattus norvegicus*, the brown rat, and *Rattus rattus*, the black, have their origins in Asia; the former in Mongolia or China, the latter farther south in what is now Malaysia. Both began their worldwide spread as humans did, following trade routes. The black rat was first – evidence of *Rattus rattus* habitation in southern Europe dates back to between the fourth and second centuries BC, followed by *Rattus norvegicus* (who, in spite of the name, don't have anything to do with Norway) although the exact date is still uncertain. From maps of the spread of early rat populations, it's possible to trace the history of worldwide trade as rat travel mirrored the bold, innovative movement of humans across the globe, following river and sea routes, man and *Rattus*, pioneering travellers together.[9]

* * *

The extended drought has me carrying every drop of water saved from the bath or shower into the parched garden in buckets. Among the tiny tomato seedlings, which I am desperately nurturing, I find a beautiful little visitor, a chameleon in luminous greens and yellows.

*   *   *

I have devoted a separate chapter to the topic of wild fish and fishing, and it is, as you would expect, concerned largely with fishing in remote, far-flung places. But it may surprise many readers to know that a city like Cape Town offers its own share of 'wild' fish. I am not referring to the coastline, which I discussed earlier as a liminal zone or seam, but rather mainly to freshwater fishing.

There is a tradition among fly fishers that if you wish to conceal the location of the hotspot at which you had spectacular fishing, you refer to it as Stream X. This tradition is the source of the name of Craig Thom's fly fishing shop in Milnerton, which is impossible to find if you do not know what you are looking for, and is identified only by a small x on the gate post. Following this tradition, but perhaps not having exercised his mental faculties sufficiently, a local angler recently posted a picture of himself on social media holding a good-sized trout, with the caption that it had been caught at Stream X. He had not considered that in the background was a large sign saying Pick n Pay, which identified it for all who live in the vicinity as that section of the Lourens River that runs past the Old Bridge Tavern, under Main Road, and past the Pick n Pay where many of us buy our groceries. We call it the Pick n Pay beat. With the recent drought, the water on the supermarket side of the bridge is barely ankle deep, but on the Old Bridge Tavern side there are some deep pools. If you are fishing up this beat, and can resist the lure of 26 different beers on tap in the pub, there are usually a couple of good fish to be had from these pools, and it is not uncommon to sit drinking a pint at a table in the beer garden and watch trout rising in deepest suburbia, in a stream wedged between the Mediclinic, an old-age residential development and the offices of a firm of accountants. Family friend and Protea angler Nick van Rensburg hooked a very large trout on this beat and, since he had left his net at home, had to persuade the homeless people who were washing their clothes in the river downstream to lend him their bucket to land it.

As the combined presence of the fly fisher and the homeless people suggests, the frequently conflicting demands of green and brown agendas play themselves out starkly in this locale. The Lourens is one of the only rivers in South Africa that is protected by environmental legislation from source to sea, though the foetid condition of the lower sections suggests that the protection is not worth the paper on which it is printed. Where it passes under what is known as the Old Bridge, it actually runs beneath two bridges. The original one, a beautiful stone structure, was completed in 1845 to provide access to wagons heading up Sir Lowry's Pass. That is reserved only for foot traffic now, as a new bridge has been constructed next to the old one for a wide double-lane road. This bridge has a number of pipes that are frequently dry, as they are designed to cope with the river in spate, and they are the semi-permanent dwellings of what seems to be a group of 35 or so people. The local council makes frequent attempts to move them, but they always return on the same day to re-establish themselves. They were not there when I first moved to this town almost ten years ago, but the vegetation along the bridge is now permanently covered in washing, and the river and its banks downstream are choked with rubbish. I experience powerfully contradictory emotions whenever I cross the bridge. I know their presence is destroying the ecosystem of the river, and contaminating the water with E. coli, and I know that a flood will seriously endanger their lives; but a forced removal in the name of ecology, especially in view of apartheid's history of brutal forced removals, is hard to contemplate.

Earlier I quoted Woolfson's comment that '[i]n one city, there are more cities than we know, hidden cities inhabited by those with whom we share everything we rely on'. In this respect, we could talk of 'fish Cape Town' or 'fishing Cape Town', which – leaving aside the vast expanses of ocean surrounding the city – comprises rivers with carp and barbel, numerous ponds with bass and bluegill, streams and small stillwaters containing trout, shallow estuaries with garrick and mullet, tilapia in small farm impoundments, water features on golf courses containing bass, tilapia and carp, pans containing the same, and so on: all parts of local ecosystems (some for worse, rather than better), and 'available' for the intrepid angler (not all of it 'legally'). And alongside the fish are coots, dabchicks, flamingoes, herons, ducks, spoonbills, sacred ibises, cormorants, and a whole host of insect, crustacean and amphibian life. The Cape Town fishing author and guide Sean Mills has developed something of a reputation as a specialist 'urban angler'.

* * *

The insistent chank-chanking of guinea fowl from the upper reaches of the oak tree tells me that one of the Cape eagle owls has made a rare diurnal visit to our garden. Usually their presence is signalled by a Whoo-Whoo call and response between male and female from the frigid darkness of a mid-winter night. Explosions of feathers and the occasional red glob of heart or lung beneath the trees by the garden shed tell of a recent visit by the African goshawk, and the demise of another dove. The agitated chattering of squirrels usually means the gymnogene is visiting. Above bird baths, bordered flower beds and paved driveways, predator and prey fight it out unchecked.

* * *

In a chapter in *Feral*, entitled 'The Never-Spotted Leopard',[10] George Monbiot discusses the story of the 'Pembrokeshire Panther', a large mythical cat, described as 'huge, jet-black [and] glossy' by the several people who claim to have seen it. He points out that there are many such beasts associated with different boroughs and villages in the UK, and that 'roughly 2000 people [. . .] see a big cat in the wild in Britain every year'.[11] The belief in secretive beasts seems to be an almost universal characteristic of human societies, and may have many bases ranging from spirituality (the large snakes associated with rivers in African societies), to deep human desires for wildness (as Monbiot suggests), or perhaps a profound existential awareness of our pathetic physical frailty as humans (anyone who has ever swum across a deep river pool will likely admit to lurking fears as to what is below one, even if one knows rationally exactly what species inhabit that body of water). In a thoughtful exploration of the topic, Monbiot dismisses the actual existence of creatures such as the Pembrokeshire Panther on the basis of lack of any evidence, but he does not dismiss the sightings; he simply argues for errors of perception.

But the coexistence of wild cats and humans need not be relegated to the realms of fantasy or mistaken identity. In place of the Pembrokeshire Panther, we could introduce the Hilton leopard, the Gordon's Bay leopard, or the Bellville caracal.

Driving home one night from Crossways pub in Hilton, just outside Pietermaritzburg in KwaZulu-Natal, some years ago a driver reported seeing a leopard zig-zagging in the road in front of his car. His account was met with the same scepticism as Oom Schalk Lourens's account of the leopard lying down alongside him under a thorn tree in the Herman Charles Bosman story 'In the Withaak's Shade', and all the expected questions were asked about his sobriety at the time. And then someone else saw it, and he was exonerated. There was indeed a leopard living in close proximity with the genteel Midlands opulence of Hilton.

Leopards are known for their ability to live very near to human habitation without being detected, and there are several leopards in the Helderberg region of the Western Cape, images of which are captured on motion-sensitive cameras. Still, when one recently took up temporary residence on the patio of a Gordon's Bay holiday home in the absence of the owners, the story made the national media. *Die Burger* carried photographs of the leopard, stalking the edge of the swimming pool, curled up next to a stack of braai wood, and wandering around amid the patio furniture.[12]

Central Bellville is a congested, grimy, industrialised urban space. It is home, among other things, to the University of the Western Cape, established in its location by the apartheid state as a place 'apart' to educate Coloured school teachers, clerks and administrators. It has long since thrown off that mantle, and the campus has been attractively landscaped, but it still nestles uncomfortably between low-cost housing developments, a Spoornet shunting yard and factories. Part of the campus is a small fynbos reserve, and – almost inexplicably, a stone's throw from a concrete yard of shipping containers and locomotives – a caracal has taken up residence there. It seems to have taken a particular fancy to the infra-red camera set up to track the movements of the nocturnal residents of the reserve, as it likes to rub up against it.

But the paradoxes and contradictions of urban wildness are perhaps best exemplified in the cats so many people keep as pets. *The Secret Lives of Cats* is a documentary produced in 2014 by National Geographic Wild. As its publicity material suggests, it will certainly cause many domestic cat owners to look at their 'pets' differently. 'Not my Fluffy!' would be the response in many cases. The documentary refers to the domestic cat as having 'the heart of a hunter', with 'wildness still cours[ing] through his veins'; it is a 'real life Jekyll and Hyde'.

At the time the film was made, there were estimated to be one million cats in the US.[13] There are 38 species of cats in the wild, including the small wildcat, which is believed to be the ancestor of domestic cats. Cats were first domesticated about 3 000 years ago. Their associations with humans have historically been contradictory in that they were worshipped in Egypt as a source of life, but also burned at the stake in Medieval Europe as companions of the devil or as witches' familiars. They have a long history of living with humans as 'barn cats', which served to keep rats at bay, and in this role their very 'wildness' was key to their 'domestication'. The documentary observes that human-cat relations are some of the most complex in nature, and that cats are more popular than dogs as pets in the Western world (one could note here the explosion of images of cats on social media).

There are as many as 50 breeds of domestic cats, and one female cat can produce up to 30 kittens per annum. It is estimated that there are twice as many kittens born daily in the US as humans. Cats become feral so easily because they are adept hunters, and much of their history with humans as defence against rodents means that this ability was prized and bred into the lineage. Their ability to climb means that they are not restricted to the ground, and can follow prey up trees or take to the branches to escape predators such as dogs. They can run at speeds of up to 48 kilometres per hour, and their highly flexible backbone means that they are very acrobatic. They are designed for nocturnal activity (hence the fact that most 'owners' see the somnolent ball of fluff, and not the top predator), with very good eyesight, including night vision, whiskers, sharp hearing, and rotational ears that can locate the source of sounds. They also have an uncanny sense of balance.

Cats can survive extreme climatic conditions. There are colonies of feral cats on Antarctic islands, which experience sub-zero temperatures, on volcanic islands, and even in the extreme heat of the Australian Outback. In some US cities, there are colonies of feral cats numbering between 75 and 100 animals, and in many cases they are too wild to be captured for spaying or neutering.

Even though they may be well fed, domestic cats impact on bird and animal populations; in fact their being well looked after means that they are in fine physical form to hunt. There are in the region of seven million cats in the UK. To try to estimate the impact of domestic cats on local wildlife, the British Mammal Society undertook a survey called 'Look What the Cat Brought in', involving 750 cats and recording what kills they brought in over a period of time. Placing a bell on the cat's collar apparently makes little or no difference to their killing ability, although the colouration of the cat does, with white cats having caught the least prey (although the top killer in the study, a cat called Missy, was black and white). The top prey items, in order, were mice, voles and shrews. When quantified, the scale of the predation is starkly evident. In one week, Missy killed five rabbits, seventeen shrews, eleven mice and seven birds. In the documentary, she looks like any other domestic cat, and her owners appear humorously indulgent of their pet's behaviour.

The impact of feral cats can be seen in a wide range of contexts, but the case of Australia is particularly striking. Cats arrived in Australia about 500 years ago, but only expanded their presence across the continent about 100 years ago with the spread of rabbits. In the space of one century, they have pushed local wildlife in the Australian Outback to margins and precarious numbers. In the absence of traditional prey species, cats have now become a significant prey species for Aboriginal people.

The Secret Lives of Cats concludes with the sobering statement that in the 60 minutes it takes to watch the documentary, 100 000 mammals and 30 000 birds have been killed by cats in the US.

A project by Sharon George of the Fitzpatrick Institute at the University of Cape Town produced similar findings to those of the UK documentary.[14] In 168 hours, one inner-city cat covered 42 hectares, travelling several kilometres per day. Several cats averaged nearly a kilometre per hour in distance. Based on the records of animals, birds and insects killed by the cats during the study, and the numbers of domestic cats in the greater Cape Town region, the study estimates that around 3.6 million creatures are killed by cats per annum. Projected to the national level, that increases to 500 million, and globally to 9.5 billion.[15]

At the outset, *The Secret Lives of Cats* says that it seeks to explore 'the paradox of the cat', the fluffy domestic companion cum top predator. It is a paradox that Ted Hughes captures so well in the poem 'Esther's Tomcat'.[16] The daytime cat 'lies stretched flat / As an old rough mat'. Barely a sign of life or animality is evident: there is 'no mouth and no eyes'; and he lies 'Like a bundle of old rope and iron'. But with the coming of dusk, 'Then reappear / his eyes', and next his mouth 'yawns wide red, / Fangs fine as a lady's needle and bright'. The metamorphosis is completed as the 'old rough mat' transforms into a killer of deep mythology, which 'sprang at a mounted knight / Locked around his neck like a trap of hooks', hundreds of years ago. The tomcat 'still / Grallochs odd dogs on the quiet', the archaic Gaelic term for disembowelling signalling its ancientness, its existence out of time, its eternal nature. It 'Will take the head clean off your simple pullet'. The skin, previously like a shabby carpet, is now impervious to 'gunshot fired point-blank', as the tomcat lords it over the nocturnal world of sleeping humans. Over a century before Hughes's poem, Henry David Thoreau had noted that 'the most domestic cat, which has lain on a rug all her days, appears quite at home in the woods, and, by her sly and stealthy behaviour, proves herself more native there than the regular inhabitants'.[17]

*　*　*

I round a corner of my house by the washing line, and one of life's greatest mysteries is inexplicably answered: a black sparrow hawk takes to the air dangling one lone sock from its talons.

*　*　*

Gardens – whether rudimentary or elaborate, productive or decorative – punctuate most urban spaces, and provide the habitats in which many wild urban lives are lived. The very fertility of the soil is a matter (literally) of wildness, and Woolfson quotes Gary Snyder's comment that '[l]ife is not just diurnal and a property of large interesting vertebrates, it is also nocturnal, anaerobic, cannibalistic, microscopic, digestive, fermentative, cooking away in the warm dark'.[18]

Gardens involve simultaneously nurturing the wild and keeping it at bay. By definition, their existence relies on the wild processes of germination, flowering, fruiting, growing. But they are also zones in which the wild is disciplined: 'weeds' are removed; trees are pruned; lawns are mown; shrubs are 'cut back'; seedlings 'thinned out'; dead annuals pulled out; and so on. Gardens may be resources for herbs, fruit, vegetables or flowers, and they may have 'wilder' and more 'ordered' sections. But they are also liminal spaces, which mediate between indoors and the world 'out there'. As I have mentioned previously, in many African societies, especially in rural areas, the notion of a 'garden' surrounding a homestead would be seen as bizarre. Instead, the area around the dwelling comprises bare earth that is kept scrupulously clean by frequent sweeping, thus making it easy to spot any unwanted intruders in the form of snakes, rodents or insects that might wish to approach the home. Suburban gardens differ across the globe, and also according to the inclinations, energies and budgets of the owners, but most seek to create some kind of more 'natural' space, outside of the hard lines of the building, either as transitional zones into the interior, or for leisure activities outside in slightly less ordered, more 'natural' ('wilder'?) surrounds. 'Gardens' in high-income areas may, however, become little more than fortified zones surrounded by electrified fencing or walls topped with razor wire to keep (wild?) intruders at bay.

Gardeners may seek to 'protect' their gardens from drought or excessive heat, by watering, mulching, erecting shade cloth and the like, but gardens are finally also as subject to the 'wildness' of weather and season as the rest of the globe. Cities, too, are at the mercy of the weather. A heavy snowfall changes not just the landscape itself, but the way in which we respond to it bodily, as Woolfson captures so well:

> Every day the snow seems still more churned and frozen. A new pavement landscape appears: a fresh set of ice ditches, ridges and mounds form and freeze. Everyone has to pay close attention to the elemental business of putting one foot in front of the other. Walking for a long time in snow seems to affect different muscles and makes me aware of small alterations and tensions in the legs and hips, a bit like the effects of walking for a long time on sand.[19]

On a far larger scale, in the face of the cataclysmic storms that sweep various parts of the globe with disturbing regularity, one is confronted with a scale of wildness that dispels utterly the illusion that human ingenuity is in control, and which vanquishes any hubristic sense of our mastery as a species, with roofs ripped off buildings, walls collapsing, roads flooded, cars tossed around like toys, homes destroyed, airports closed and the roads barely traversable (and, as ever, the poorest of the poor bear the brunt). The brute power and engineering feat of bulldozer or front-end loader are completely ineffectual against millions of kilolitres of sludge and debris. As frail mammals, we feel – and are – puny.

* * *

We are back in our suburban home, after an afternoon of pursuing trout. There were thrushes and robins scurrying through the undergrowth, an owl hooting intermittently into the warm afternoon air, a black duck breaking cover as we approached the water, a jackal buzzard lifting itself into the air on rounded wings, an owl pellet in the grass as we approached the wooden bridge, and of course the water and its trout. This evening we feel for a time that we have been touched by something . . .

* * *

I want to stress, in conclusion, that in referring to cities as zones that contain 'wildness', I do not mean in any way to belittle their (and hence our) own efforts to obliterate that very thing. Woolfson quotes Sybille Bedford's notion of 'cities' as 'heat islands', which fundamentally transform temperature, climate and environment, by the nature of their construction:

> We may be northern and cold [she is writing from Aberdeen] but if we're in a city, we're in a heat island, a place where temperatures are higher than in the surrounding countryside by as much as eleven degrees. We're heat islands because of the way we build and materials we use to build – our concrete, glass and stone. It's our roads, our roofs, our pavements, our dark, absorbent surfaces, our paving of gardens, our cutting down of trees. It's the way we live too, our heating or our cooling, our determined use of cars. We're bounded by the heat we generate, our energy rising to the skies.[20]

The wild in cities is testament almost entirely to its own resilience, and only much more recently and minutely to active human promotion and protection.

7

The Wild and the Farmed

Wyoming. Now that's an organic food store.
　　　　　— Advertisement for Wolverine hunting boots.
　　　　　　　　Field and Stream magazine

Man has been defined as a rational animal, a laughing
animal, a tool-using animal and so on. We would be
touching upon a deep truth about him, however, if we
called him a cooking animal.
　　　　　　— Martin Versfeld, *Food for Thought*

I began to see how the history of food is geography,
immigration, culture, urban planning, science, technology,
education, real estate, economics: the history of the city
itself.
　　　　　　— Robin Shulman, *Eat the City*

Agricultural shows are important, and housewives compete
with tarts . . .
　　　　　— Betsie Roodt, *Betsie Rood's 101 Traditional
　　　　　　　　South African Recipes*

I discussed in Chapter 2 gradations of wildness, of wildness as
variant quality rather than absolute category, drawing on the work of
Gary Snyder and George Monbiot. Here I consider the argument

114

in relation to the human activities of cooking, food production or collection, in particular, as my chapter title suggests, what has frequently been seen as a fundamental distinction in understanding human relationships with our environments: that between the 'wild' and the 'farmed'. My point is not just to suggest that the 'wild' and 'farmed' are categories far more complex and unstable than we may initially assume, but in so doing to direct attention to the crucial questions (if we take the biological and ethical aspects of (re)wilding seriously) of where our food comes from, and what practices are involved in its production or collection.

I begin with two illustrations that unsettle customary understandings of the distinction between the 'wild' and the 'farmed'. The first relates to legislation about indigenous and alien species. Part of the difficulty for the Department of Environmental Affairs in producing legislation about species such as trout in South Africa is the fact that the species exists in 'wild', self-sustaining populations in rivers, at one end of the spectrum, and in fish farms where they are bred for consumption, rather like pigs or chickens, at the other. And to complicate matters, farmed fish can, say through flood or the bursting of dams, rapidly become 'feral' or 'wild', as has happened with pigs, dogs and cats in various parts of the world. The assumptions of current legislation in South Africa are that fish such as trout are 'wild' (even if they are being farmed in porta-pools in a city with no chance of escaping, as they are in Grahamstown, for example), and so – as fish famers lament – environmental restrictions are placed on them that are far more stringent than those applied to any other farming practice.

The second relates to legislative understandings of ownership. With the rapid expansion of human populations in the last 200 years, competition for food has increased substantially, whether 'wild' or 'farmed'. When the question of human need (or greed) factors into understandings of wild environments, things become decidedly complicated, as questions of rights and ownership enter the fray in ways very similar to those regarding 'farmed' crops or livestock. As I have pointed out previously, in a strange contradiction, 'wild' animals can also be 'owned' by humans. This may not be difficult to understand where the animals reside on land owned and fenced by an individual or company, but even when this is not the case – for example with marine fish – ownership may still be asserted and legislated. We are all aware of the legislation governing the catching of marine species, both at national and international levels, sadly too often honoured in the breach. Declaring the rights of a country to exclude any foreign fishing activity within a certain distance of its coastline is nothing less than claiming ownership of all that lives in, or migrates though, those areas of sea.

But the same happens at more local levels, where things become perhaps more interesting. A chance find (literally a couple of loose sheets that appeared to have fallen off the photocopy machine at work) pointed me to a fascinating historical episode. It was a report from the Appellate Division in Bloemfontein, dated '1921. March 15, 17' in the case of *Van Breda and Others v Jacobs and Others*, in which two groups of fishermen were disputing ownership of a catch of fish. Both groups operated in boats off Simonstown, and on 9 September 1919 spotted a large shoal of fish travelling from Simonstown towards Glencairn. Both boats made unsuccessful attempts to intercept the shoal off Klein Fish Hoek, and then both proceeded to Kamartje to try again. The plaintiffs reached the destination first, and laid out their line, while the defendants were freeing their boat from some wire from a stranded ship. On finally arriving, the defendants laid their line between the approaching shoal and the line of the plaintiffs, and caught the entire shoal. Citing the custom among fishermen that the second arriving boat may not set up line or net between the first boat and the fish, the plaintiffs argued, successfully, that the defendants had 'taken their fish', and that they should be compensated financially for their 'loss', ascribing precise monetary value to that loss. They insisted that they 'owned' the shoal of fish that they had not caught, and the court agreed.

* * *

My brother hands me two large packets. Even through the plastic I can smell the warm aromas of coriander and black pepper. Biltong and dry wors from his last hunting trip. I am no hunter myself, though in the past I have shot doves, pigeons and even a guinea fowl, and gutted and dressed them for cooking without squeamishness. On a schoolboy visit to a friend's game farm, I watched his father drop an impala about 80 metres from us with a rifle. I recall the percussive noise of the shot, the congealing blood, and the glazing eyes of the dying antelope – it was not an experience I would choose to repeat. But in an important sense I have to admit that my brother is more honest than I am, or many others are, in accepting responsibility for killing what he eats, for being present in the moment of death. Supermarket packaging provides a flimsy covering over of the blood and violence of eating meat. Animals undoubtedly know fear and distress. What is more humane: a bullet out of the blue or an overcrowded truck ride to the abattoir?

* * *

In his study, *Guns, Germs and Steel: A Short History of Everybody for the Last 13 000 Years*, Jared Diamond presents an account of human interaction with animal and plant species in the 13 000 years since the last ice age, the years in which humans as species made their most rapid developments. Particularly striking in the study is just how recent the development of human societies has been in some areas of the globe. If Diamond's argument for the occupation of New Zealand as recently as only 1 000 years ago is correct, then the arguments of some scientists that presence in an area for several thousand years is insufficient for something to be called indigenous seem extremely unhelpful, particularly because the concept of indigeneity within most arguments about conservation is that which guarantees value and belonging, rather like citizenship for non-humans.

Diamond also complicates arguments about 'wild' and 'farmed' food, and supposedly absolute distinctions between the hunter-gatherers, herders and farmers who collected or cultivated it. The 'development' from hunter-gatherer to agricultural existence, and hence from 'nomadic' to 'settled' life, from 'subsistence' on 'wild' food to the 'cultivation' of grains, animals and vegetables to the point of 'surplus', has been fundamental to many accounts of human evolution and the development of 'culture', even Culture (with its 'High Art' associations). But Diamond points out that many societies historically had mixed-mode food economies, that some farming societies reverted to hunting and gathering for periods, and then went back to farming, and that even hunter-gatherers sometimes 'cultivated' their wild stocks of food by trimming back trees to promote new growth, replanting after harvesting, setting fires to promote the growth of new shoots (Australian Aboriginal 'firestick farming'), damming or redirecting watercourses to create 'fish farms', and so on.[1]

Our own diets range across the categories of the 'wild' and the 'farmed', sometimes in ways of which we may not be aware, unless we have learned – whether for reasons of ethics, health or sustainability – to understand the sources of what we eat. Recent studies suggest that at least 50 per cent of fish consumed worldwide is farmed,[2] and that commercial fishing has destroyed about 95 per cent of the wild stocks of fish such as dusky kabeljou – a highly prized table and angling fish. As I have noted though, fish farmers in South Africa complain that the legislation concerning the commercial utilisation of fish species is based entirely on the notion of 'wild' stocks, and places on aquaculture 'more stringent demands than [it does on] the wild-caught industry'.[3]

Some years ago, when pondering these issues while writing *Are Trout South African?*, I attended an aquaculture course, held at what was then still the Jonkershoek hatchery. Several of those who had signed up were especially interested in the possibilities of farming tilapia, either as a business venture or as a development initiative to alleviate problems of food insecurity. One of the presenters pointed out that a key problem with tilapia farming in South Africa is that the average water temperatures are too cold for optimal growth, and so the water has to be warmed to get anything like a reasonable time to harvest. This makes the fish expensive to raise, and unless one is able to get them into the high end of the market as a desirable table species (as has happened with tilapia through marketing campaigns in the US) one will battle to sell them at a profit. He remarked that at the lower end of the market, farmed tilapia would be competing against (wild-caught) canned pilchards, which are widely consumed in South Africa. Pilchards are high in protein and omega-3 fatty acids, and a 400-gram tin in a supermarket costs about the same as a loaf of bread. This is not to suggest that farming tilapia could not be an important supplement to subsistence farming. It can and is, especially if the fish are fed on products that would otherwise be waste (including vegetable peelings and chicken litter), and also if the time to harvest is not a major issue.

Attempts were also made in South Africa to farm catfish, and market them as either a staple fish or as a high-end product ('freshwater kingklip' was one ambitious attempt), but they have never seemed to find acceptance from consumers, though many African people catch and eat them. The success story of South African aquaculture is trout, which do well in our high-altitude regions, and fit into the very top end of the market. There are also current (and controversial) experiments with farming trout and salmon in cages in the sea off our coastline, as well as some aquaculture projects farming kabeljou. Again, these are aimed at the top end of the consumer market.

So, in terms of widespread consumption in South Africa, farmed freshwater fish, whether tilapia or other species, are edged out of the market place by tinned wild fish. But their taste is apparently not to the liking of all fish eaters who reside within our borders. I recently attended a seminar at the Institute for Poverty, Land and Agrarian Studies (PLAAS) at my university on 'The Trade of Fish and Fish Products in the Southern African Corridor' by M. Hara, T. Jimu, E.E. Ndofor Epa, E. Kaunda and S. Chimatiro. It opened my eyes to regional and urban food networks of which I had been previously unaware.

The paper focused on a range of fish products brought into South Africa mostly to cater for people from the African diaspora living in the country (of which there are estimated to be about five million). The products range from, for example, small sun-dried fish such as kapenta (from Zimbabwe and Mozambique) or usipa (from Malawi), crayfish (from Cameroon), smoked dried catfish (from West Africa, Zambia and

Malawi), stockfish heads (from Norwegian cod re-exported through Nigeria), fresh chambo (from Malawi), bream (from Zimbabwe), and many others. These are all 'wild' species. Chinese farmed tilapia is also imported, as this is cheaper than locally produced fish. Most of the fish products are dried, frozen or smoked, and are sold in specialist shops in the Western Cape, or from market stalls in Gauteng. Though the fish are expensive in comparison with South African products, those living far from their countries of birth clearly value the 'tastes of home' – of the fish they would have caught themselves or bought from markets – enough to pay a high price for them. Out of the subcontinent, in the reverse direction, go tinned sardines (from South Africa), horse mackerel (from Namibia) and Chinese farmed tilapia.

The trade networks are informal, but efficient. Some of the fish products from Nigeria and Cameroon are carried on container ships, but most of the fish products come into South Africa by bus and truck, brought in mainly by travellers and informal traders, who exploit the regulation that goods in amounts of less than 500 kilograms are not subject to import taxes. The products being brought in are also often unknown to customs and border control officials, who are then not sure how to assess their value, fitness for human consumption, and so on, so there is typically little control, though traders do complain about victimisation and harassment from officials. So a complex, but largely unregulated, network of capture, processing, transporting and selling keeps diasporic communities in South Africa supplied with a wide array of fish products that are fundamental to their identities and cuisines, something that multiple studies of exiled communities have shown take on heightened significance in contexts of displacement.

Aquaculture in Africa is widely practised in Egypt, and also in Morocco, but the bulk of the diasporic trade is in species harvested from wild populations, as noted above. The presenter of the seminar noted wryly that while the importation of Chinese farmed tilapia may harm the local fish trade, already threatened in many African countries by the spread of South African supermarkets into the Southern African Development Community region and the ready supply of South African tinned fish, Chinese tilapia may potentially also save many indigenous fish species from over-exploitation. (One can note here the irony that, despite its current highly invasive status, the importation of the black wattle into South Africa for firewood prevented the destruction of some indigenous forest areas.) What remains to be seen is whether the transportation of fish products 'from elsewhere' may have any effect on local environments. All of these products, and any pathogens they carry, will inevitably end up in the waterways in the form of sewerage.

*   *   *

Collecting mussels is a simple joy. Hopping across rocks covered in barnacles and great honey-combed colonies of marine worms, looking for the largest and plumpest specimens, and levering them out of their beds while trying to do as little damage to the surrounding mussels as possible. And keeping an eye out for waves, of course, which can give you a mild dunking, or get you into real trouble if the swell is up. The right footwear is essential, as a fall can be painful and dangerous. The mussels go into a mesh bag, which will later be left in a rock pool so that they can spit out any sand they have ingested. Steamed in the battered old pot that we reserve for this purpose, with a splash of wine, a bay leaf, some black peppercorns, a halved clove of garlic and a sprinkling of herbs, they will be delicious with brown bread and a cold glass of white wine at lunchtime. There is something primally satisfying about collecting your own food.

With musselling come the other pleasures: bright anemones in blues and pinks; gardens of sea urchins in purples and maroons; cockles with their pitted white trapdoors; periwinkles with a swirl of pink on the base of the shell; the occasional feathery sea slug or brightly armoured star fish; and sometimes an elusive octopus, flicking a speculative tentacle from beneath a rock.

Once the mussels are counted, cleaned and left to soak, it's time to collect some redbait. The leathery pods betray their presence with twin jets of water from their teats as the water retreats in runnels around the bases of the rocks. A sharp knife cut reveals firm orange flesh, with a fresh salty smell – far removed from the pale, stinking flesh of those pods that wash up on the beach after heavy seas. The redbait is for the afternoon, when the high spring tide floods into areas ordinarily not much more than rock pools, and the fish come in to forage places they can seldom otherwise reach. Our tackle for this fishing is as light as we can get away with. Years of fishing this spot tell us that if the blacktail come in today, they will be holding in a hole not much bigger than a garden pond, a cast of about 20 metres away. There are usually some big fish among them, darkened with age, with heads and jaws more like those of steenbras than the beaky appearance of their younger siblings. If the redbait is fresh, your cast is pinpoint, and the fish are there, the result is almost instantaneous, as the blacktail grab the bait with a force that pulls the rod tip down fiercely. Once hooked, they generally take the channel back out to sea. If they go either left or right, they will snap you off on the reef. On the big ones my success rate is about 50 per cent. Once you have the fish inside the reef, it is a mighty battle as it turns broadside and you are pulling against the strength and breadth of its body. When you have them close, you need a pushing wave to lift them onto the rock. Many times I unhook them as carefully as I can, and love their quick dart away into the surging water. They disappear so quickly that I wonder if I have dreamt them. Sometimes I tap them over the head and treat myself to a dinner of fish so fresh that it might still be swimming.

And then, just in case you are starting to feel self-satisfied about having cracked the code, the tide will be perfect, the swell will be pushing just enough, the redbait will be plump, fresh and enticingly attached to the hook, your casting will be pinpoint, and you will catch absolutely nothing, not even a pesky, bait-stealing klipfish.

* * *

'Wild' food is a resource, still freely available if one knows what to look for, but requiring scrupulous care in its collection and use. It has been exploited for millennia by rural people across the globe, but it is surprisingly available to urban dwellers as well, as a whole new generation of self-styled 'urban foragers' have demonstrated in recent years. I referred to Kobus van der Merwe's restaurant Oep ve Koep in the opening chapter of this book, and how its commitment to local, sustainable food blurs the categories of the wild and the farmed, as well as the indigenous and the alien. I want to take the discussion of his cooking and foraging philosophy further. These are explained in some detail in Van der Merwe's book, *Strandveldfood: A West Coast Odyssey*, with exquisite photography by Jac de Villiers. Van der Merwe has since moved to new premises in Paternoster and opened another restaurant called Wolfgat, more of which below, but the culinary and collecting journey starts with Oep ve Koep, which still operates as a general store-cum-coffee shop.

Oep ve Koep is housed in 'an old shark-liver oil factory building dating back to 1923 [which] was first turned into a West Coast general store in the 1980s'.[4] Van der Merwe had previously worked as web editor for the restaurant guide *Eat Out*, before frustration with an office environment led him to pack that in and head up the West Coast to run the kitchen at his parents' 'small-town-shop-meets-general-dealer-and-bakery [. . .] Die Winkel op Paternoster (or Oep ve Koep as the locals know it)'.[5] A truly remarkable journey into wild food and localised cuisine ensued.

Collecting wild food is something Van der Merwe remembers clearly from childhood, and he talks of 'fond memories of picking mussels, *siffies* (abalone) and periwinkles off the rocks during school holidays at Jongensfontein in the Southern Cape and helping my grandmother collect seaweed to make jelly that she flavoured with port'.[6] It was not just the sea that provided such rich possibilities, as he recalls 'collecting wild cucumber (*Cucumis africanus*) and karkoere (*Citrillus lanatus*) on the family farm outside Kuruman in the Northern Cape', and his grandfather showing him 'the *!nabba* or Kalahari truffle on a neighbouring farm close to the Botswana border'. He goes on, 'Of course, there were the suburban "foraging" outings for loquats and *moerbeie* (mulberries – berries for ourselves, leaves for our silkworms), and the walks in the pine plantations, eating pine kernels straight from the cone.'[7]

The supplementing of their own produce by harvesting wild sources by farmers, described above, finds resonance in an unpublished novel (found by my colleague, Julia Martin) about a late nineteenth-century English farmer's wife in what is now KwaZulu-Natal who describes in great detail collecting leaves and berries from the veld for both culinary and medicinal purposes. Despite her fairly recent translocation from the northern hemisphere, the extent of the knowledge of South African plants that she reveals is astonishing.[8]

Harvesting mussels, crabs, fish, perlemoen, crayfish and alikreukels has been part of my life for as long as I can remember, and something that still brings me enormous pleasure. The recipe books *Free from the Sea* and *More from the Sea* were standard reference books in our Eastern Cape cottage.[9] I suspect a commitment to and respect for the sea as a resource lay behind the fact that my father could never understand the idea of catching and releasing fish once you had caught your limit, nor of fishing for species that you could not eat. Fishing for him was as much about catching delicious food as it was about sport, and both contributed to the pleasure and satisfaction of the pursuit.

But I also have memories of food expeditions in more unusual places. We lived for eight years in the Free State town of Welkom, and as a family would regularly head out over weekends in season to collect the mushrooms that grew in the parks and on the traffic islands. As my father had a largely urban childhood, I assume we must have drawn on my mother's experience of being brought up on a farm and knowing exactly which ones were edible and which were not (we had also lived in the more rural environment in Greytown prior to Welkom). There were beautiful big ones with dark gills and a glossy brown covering on the cap, tightly folded button mushrooms, others with delicate pink gills – but beware those with white gills or dotted white caps.

Van der Merwe's concern with wild food is also driven by an interest in the history of local cuisine. He talks of his '(healthy) obsession' with C. Louis Leipoldt, who covers indigenous wild food quite extensively in his cookery books, and notes that '[f]or research on local ingredients, I usually turn to his notes on *veldkos* first'.[10] Intriguingly, he notes in this regard, that 'Leipoldt regarded the flamingo as one of our best indigenous game birds, especially singing the praises of its breast meat'.[11] Other sources include '*Kos uit die Veldkombuis* (Betsie Rood), and the classic *Food from the Veld* (Fox and Norwood Young)'.[12]

Van der Merwe's emphasis is on presenting food that reflects the locality of what he refers to as the Strandveld,[13] and the cuisine, drawing on the associations of both 'strand' and 'veld', is 'naturalist, modern country cooking with a strong connection to the land and ocean'.[14] As well as being dedicated to locality, Van der Merwe's dishes 'reflect extreme seasonality – something that almost captures an exact moment in time, on a plate.' He says, 'Even when making something as simple as a "garden salad", it must be exactly that: a representative mix of herbs and leaves, a snapshot straight from the garden.'[15]

Those with a concern for sustainability place emphasis on the production and consumption of local foods, both as a way of supporting local communities and farmers, and reducing the carbon costs of transporting foodstuffs over long distances. For Van der Merwe, however, this is not just an environmental issue, but also one of identity and locatedness. He says that he 'started experimenting with pairing only ingredients found in a specific locale [as] the only way to truly taste a region: by nibbling on the wild leaves, roots, flowers, seeds and shoots inherently produced by the very earth that you tread on'. From these beginnings began a more adventurous journey:

> The more I explored the wild food of the area, the more I started discovering truly new tastes. I started by comparing them to conventional or known tastes, pairing wild ingredients with cultivated, wild with wild, questioning everything. And then things started to fall into place. *Mosselbank at low tide* is one of the very first dishes that came about in this way, and it will certainly remain a signature dish for many years to come. Today, many of the dishes carry names of places where I've collected specific ingredients or of inspiring locations in my immediate surrounds.[16]

As well as drawing on the culinary traditions described by Leipoldt and others, Van der Merwe's approach also looks to longer histories, including archaeological excavations of middens that reveal that early inhabitants of the region 'lived on a diet of mainly small game, a variety of shellfish and wild greens', leading him, for example, to create a dish combining these flavours, which he refers to as the 'mid-Holocenic hunter-forager's "surf and turf"'.[17] It comprises springbok, limpets, seaweeds, and various other ingredients, and is delicious, as I can vouch for from personal experience, having eaten it at Oep ve Koep.

With fond memories of dining at Oep ve Koep, an even keener interest in wild and foraged food developed in the process of researching this book, and having heard reports that Van der Merwe was taking his cuisine to new levels, I headed up the coast to visit his new restaurant, Wolfgat, also in Paternoster. 'New' may in some ways be something of a misnomer, as the building in which it is housed is 130 years old. It commands a vista of the bay at Paternoster that is simply breathtaking. Having opened its doors in August 2016, it is fast making a name for itself far beyond the shores of South Africa. A note on the wine list indicates that the building is just above the famous Wolfgat cave, a 'site of immense archaeological and geological interest'. An initial archaeological survey apparently 'found ceramic remains and sheep bones dating from some time in the last 2 000 years', as well as 'marine shell, ostrich eggshell, ceramics, beads and stone artefacts'. The panoramic view, the long history of the building, and the site being steeped in the food and foraging history of Paternoster – all these suggest an extraordinarily appropriate location for such a restaurant. As we await our menus, we watch a fisherman just off the beach back-paddling his fishing boat with practised skill to the end of his net line, and flipping fish after fish quickly into the boat behind him.

Lunch is a seven-course taster menu, which showcases locally produced and foraged foods, almost all seasonal, as the menu changes according to what is available. The foraged plants are identified by both local common names and scientific genus and species. The menu on my visit was:

Strandveld snacks
Swartmossel, elandsvyl / Soutslaai, waatlemoen, angel fish
Carpobrotus quadrifidus, Mesembryanthemum guerichianum

Tjokka
Wild garlic masala, nartjie, slangbessies
Tulbaghia violaceae, Lycium ferocissimum

Heerenbone
Strandveld greens
Trachyandra cilliata, Spalmanthus canaliculatus, Tetragonia decumbens

West Coast oyster
Quince, kelp / Pomelo, dune celery
Laminaria pallida, Dasispermum suffruticosum

Saldanha Bay mussels
Klipkombers, Sandveld patat
Pyropia capensis

Springbok
Brakvygie, snoekkuite
Aptenia cordifolia

Wild sage
Amasi, sage ash, nectarine
Salvia Africana-lutea

The combination of ingredients – in terms of taste, colour, texture, aroma – is extraordinary, and Van der Merwe creates from local ingredients, many of which would be dismissed as 'weeds', a dining experience of the highest order. Those 'weeds', as well as the mussels, oysters, fish and venison, have of course fed the inhabitants of this region for many centuries, and in some cases still do.

I indicate to Van der Merwe that I have been trying to contact him to talk about his work and my own research, but it seems the vagaries of cyberspace have not connected us. But we exchange contact details, and a couple of weeks later I meet up with him on a weekday as the lunchtime service is coming to an end. I sit at a table looking out over the inside section of the restaurant. The diners are out on the verandah in this beautiful summer weather. I take in the bottles of pickled leaves and stems that line the walls, and the view over the bay of white sand and crisp blue water, and watch the staff in the kitchen plating up the dessert. Over a cup of coffee, he and I start to talk.

Van der Merwe, whom it now seems more appropriate to call Kobus, proves to be a good and interesting conversationalist, while nevertheless keeping an attentive eye on the kitchen, the staff and the final course going out. In response to my question about why his fascination with the wild and the foraged, he indicates that he is concerned with trying to represent 'where we are'. He stresses that this approach is in stark contrast to what he refers to as 'the commodification of place', in which visitors to the West Coast, for example, are served calamari and chips under the guise of sampling local seafood cuisine (the calamari likely of Patagonian origin). But he also takes me back to his childhood, in which Leipoldt's cookery books were staples in his home, and family holidays were, as he stresses in *Strandveldfood*, extended foraging expeditions, in which every family member, regardless of age, had the requisite licence to contribute to the seafood collection.

The move to Oep ve Koep, which he describes in his book, was actually supposed to last only a year, as he admits he was imaginatively tied to Kalk Bay. But eight years later he is still in Paternoster, though he shuttles been the two on fortnightly visits. I ask about the rationale behind the new restaurant. Kobus points out that Oep ve Koep was a space that had already been created, and the cuisine somehow had to fit into that. Even with retro-fitting some of the decor, he felt it was not always an ideal match. He then moved to hosting dinners in the attic, which was previously a boat repairing facility, which suggested

the possibility of something different and more expansive, something
of a blueprint for what he later sought to achieve in Wolfgat. With the
130-year-old building, he felt he had a blank canvas, a space in which
a different story of food and place could unfold. It is a story of many
characters. Directly in front of the restaurant is the historical lookout
point, Die Krantz, from which the fishermen check sea and weather
conditions daily, and have done for decades, and which is apparently
host to an ever-changing cast of visitors throughout the day, from the
early morning coffee drinkers, the marijuana smokers, the school kids,
and the late afternoon drinkers with their large glasses of brandy and
Coke.

Kobus picks many of the plants he uses from the farm on which
he lives, which he points to somewhere across the bay from where
we are sitting. He is currently experimenting with coastal lamb from
Verlorenvlei. His own personal life of cooking, foraging and collecting
intersects with that of the restaurant; at present, he says, he is trying out
lamb recipes at home, some of which may make it onto the restaurant
menu. Part of the lamb fascination is the distinctiveness of the flavour
of animals that have been feeding on coastal vegetation. But another
crucial part is that excavations of early settlement sites of late Stone
Age peoples (including Wolfgat cave itself) indicate that lamb was a
major part of their diet. They also ate seal and used the fat to preserve
food, but, despite owning a kayak, Kobus does not appear that keen
on pursuing that line of fare. He stresses that the cuisine is local but
not fanatically hyperlocal, though he admits there are aspects of that.
The criteria for inclusion appear to be that the ingredients should have
some association, whether current or historical, with the place. The
springbok is from the Northern Cape, and the !nara oil from Namibia.
The olive oil is from the Western Cape, not imported, and the wines
are from estates nearby.

As to how he comes up with his dishes, he says that his approach is
largely intuitive, but that local, sustainable produce (mussels or oysters,
for example) should be the starring ingredients when used, and tastes

and textures sought to complement them. One of his major difficulties is that the collecting time for local plants is winter, whereas the bulk of his diner custom is in high summer. Some of the collecting seasons are very short – only about three weeks in the case of veldkool – hence the need to preserve many of them in pickles of various kinds, in which local herbs are used to provide flavouring, and the multiple bottles lining the restaurant walls. He and local food blogger Loubie Rusch apparently have an annual foraging and pickling date that coincides with the best harvest period.

Towards the end our conversation is punctuated by Kobus having to excuse himself to receive payments, and compliments, from happy diners. I notice that the accents are almost all foreign, an impression that he confirms in telling me that his clientele is mostly foreign tourists, at present. This does bring me to the question of cost. Oep ve Koep was probably in the mid-range of restaurant prices. Wolfgat may have you needing to sell off an organ (not just a kidney, but a heart), though I would acknowledge that there are some more expensive restaurants in the Western Cape. My question to Kobus in this regard is not actually about whether the food is worth the price – I can safely say it is the best meal I have ever eaten, anywhere – but whether the model of harvesting local, wild resources is sustainable at lower economic levels.

His response is twofold. On the prices at Wolfgat, he points out that the emphasis on local produce meant the need for a licence to work with abalone, the cost of which is astronomical; that taking foraged food cuisine to the next level is extremely time-consuming and labour intensive; and that they also sought to build a really good wine list of local estates to complement the food, which comes at a cost. But he admits to having a dream ('romantic', is his term) of extending the idea of foraging and collecting to the point that we would replace all the wheat fields with edible coastal and fynbos species.

But perhaps this is not just 'romantic'. In *Strandveldfood*, Kobus argues that the approach to the food, its collection and preparation is not simply a 'fad' associated with 'new primitives' or bored gastronomes seeking novel experiences:

In South Africa we are a nation of heritage foragers. To this day, a large percentage of the population still relies on wild food as part of their daily diet. On the West Coast, veldkool (*Trachyandra* sp.), black mussels, white mussels, limpets, even melkelsies (*Cynanchum africanum*) and bietou berries (*Chrysanthemoides monilifera*) have been part of daily food culture for generations.[18]

That said, he stresses: 'It is imperative to have an almost fanatical, built-in sense of responsibility towards the environment when you set out to collect food from the wild. Always pick with permission, with a conscience, and with future seasons in mind.'[19] He also cultivates many of these indigenous food plants in his garden. He insists that our old patterns of food production and consumption are no longer sustainable or ethical:

Now, more than ever, we should go back to our roots, consume cautiously and with a conscience. We should get to know our indigenous bounty, respect it, use it wisely and manage it sustainably with future generations in mind. Growing our own leaves and vegetables at home should not be a trend; it must become a way of life. We must completely rethink our obsession with meat and its by-products and ruthlessly and continuously question the what, where and how of any seafood – from prawns, to tuna, even bokkoms – that lands on our plates.[20]

The last diner has left by the time my patient interlocutor has answered the last of my questions. The staff members are in the process of closing up, and the captivating view is shut off by closing doors, as I thank Kobus for his time.

* * *

As part of a broader commitment to ethical eating and the nose-to-tail culinary philosophy, but also – if truth be told – probably to show that we are adventurous eaters undeterred by the unconventional, my wife and I decided to make sheep's trotter and bean curry. Nose-to-tail it may have been, but environmentally friendly it wasn't, as we finally found the trotters in Paarl, a round trip of 90 kilometres from our home, with the attendant carbon emissions. It would probably have been cheaper to buy fillet steak from the butchery around the corner. For reasons that I still cannot fathom, the actual making of the curry was left entirely to me. The trotters were boiled for several hours, emitting a reek of something like unwashed sheep's pelt and boiling fat (this despite the fact that the trotters had been impeccably cleaned by the butcher). The smell clung to clothing and furnishings for days. The bean curry and accompanying sauce had to be made separately, turning this into an operation that took most of the day. Several pungent hours later, I assembled the dish and presented it to my family. Their response was queasily polite, and they picked unenthusiastically at their bowls. After a suitably diplomatic time had elapsed – probably about ten minutes – I was told to give the lot to the Great Danes.

* * *

Thinking about wild and farmed food, and its collection, growth and preparation (both in ethical and culinary terms) takes one on unexpected journeys, into areas of cities one might otherwise have overlooked. I mentioned discovering an urban network of diasporic fish products in my own home city earlier, and memories of collecting mushrooms with my family in the town of Welkom. Living in Pietermaritzburg, on my trips to and from work at the University of KwaZulu-Natal, I would regularly see subsistence fishermen with handlines on the banks of the Umsunduzi River, which runs through the town, targeting carp, barbel and yellowfish (the foetid nature of that watercourse is tragic testament to their desperation), and the same is to be seen on the banks of many of the ponds, canals and rivers in the greater Cape Town area.

In a strange paradox, while the world's major cities may reflect the lowest quality of wildness as measured in conventional terms, especially as related to animals, they may have the highest quality of wildness measured rather differently – in terms of what we consume – especially if they have large immigrant communities.[21] New York is an excellent example. In her book *Eat the City*, Robin Shulman produces an alternative account of the city of New York entirely though the 'fishers, foragers, butchers, farmers, poultry minders, sugar refiners, cane cutters, winemakers, and brewers' who built it. Much of what she describes would fall conventionally into the category 'farmed' rather than 'wild', and the reader may legitimately ask why I appear to be detouring off a discussion of 'wildness'. My answer is twofold. Firstly, engaging with 'wildness' necessarily involves engaging with biological processes, and hence with the origins and sources of what we eat, whether cultivated or collected. Secondly, in the city, the 'farm' has a quality of wild 'otherness'; it is a place where things are grown and harvested, where animals are raised and killed, rather than presented in hermetically sealed vacuum packs, which may explain the insistent attempts of city authorities over many years to shut them down. 'Farmed' and 'wild' are concepts more complex, and sometimes intertwined, than we may assume.

Eat the City is partly about Shulman's own experience of New York and its ebbing and waning relationship with food production. She describes how, as a student, she noticed:

> In a few years, the far east of Fourth Street had gone from buildings to prairie to a small working farm. All through the neighborhood, wherever people could find an empty patch of ground, they were planting tomatoes, squashes, and greens, raising chickens and rabbits and turkeys and ducks, living off the fat of the urban landscape . . .[22]

But that food production has for many years been precariously at the mercy of town planners and 'developers', and she notes after some years working as a journalist in the Middle East how shocked she was to return in 2005 to see that 'little was recognizable. Gardens had been bulldozed . . . Landlords were recruiting more high-rent professionals, and a lot of my former neighbors had moved away.'[23]

In her investigations, Shulman meets 'people who grow vegetables and fruits and mushrooms, who fish and forage, who go clamming and trapping, who collect honey, who produce cheese and yogurt, who make beer, wine, hard liquor and liqueurs, who keep goats for milk, and quails, ducks and chickens for eggs, who butcher city-grown rabbits, turkeys, roosters, and pigs'. And she visits 'rooftops and basements, rivers and fish tanks, fire escapes, window ledges, warehouses, packing plants, storefronts, breweries, wineries, and community farms'.[24] Those involved comprise 'misplaced rural folk dreaming farm dreams on the subway', 'next-generation foodies forever seeking a more complex and rustic thrill from the homemade', 'people of limited means looking to save or make money', 'eccentrics obsessed with process', 'sentimentalists chasing tradition', 'parents concerned with health', 'professional artisans', 'manufacturers focusing on profit' and 'philosopher-farmers trying to build a new American urban life'.[25]

Many are immigrants who – like the diasporic inhabitants of Cape Town or Gauteng with their fish imports – are seeking to recreate the food 'of home' by cultivating it in the new city. Shulman refers to the Sholom Aleichem lullaby, which Eastern European mothers used to sing to their children more than a century ago – 'in America there will be chicken soup in the middle of the week' – noting that they 'fulfilled that promise when they arrived in the newly constructed yet already crowded tenement buildings of my neighborhood, where they kept chickens, ducks, turkeys and geese. There was squawking and clucking in hallways, apartments, basements, and narrow airshaft yards [. . .] For how long has a rooster wakened the people of Fourth Street? A hundred years? Two hundred?'[26]

Eat the City is, then, not simply an account of craft beer, boutique wines, organic vegetables and kale, consumed by new age foodies or hipsters, though that is part of the narrative, but also an account that seeks to ask: 'How did history bring us to where we are? In an enormous, overdeveloped city with millions of citizens and hundreds of years of momentum, how do people mark the landscape with their own personal hunger?'[27]

I mentioned above that there are surprisingly abundant sources of 'wild' food even for urban dwellers, but if you add to that the possibilities for 'farming', the urban landscape is even more fecund: edible plants grow wild or can be cultivated on vacant lots or in parks or gardens; rooftops can become gardens; basements and backyards can house livestock or even fish tanks; seafood can be caught from piers and waterfronts; vegetables can be grown on balconies or in pots.[28]

Shulman covers a wide range of food-related topics in relation to New York. She discusses New York's history of sugar importation and refining, and its implication in the slave trade, which provided plantation labour. She narrates the history of beer brewing in New York, right up to its current craft beer manifestations. And she recounts the historical and present wine production in the city, in particular kosher wine, and in one enchanting case wine produced from the grapes of a vine, some 15 metres or more in height, in the backyard of an otherwise unassuming double-storey house, the harvesting and pressing of which have become an extended family tradition.[29] But those are less germane to my current discussion of wildness and food. I want to focus instead on bees, vegetables, meat and fish.

Bees and the production of honey have for many years been part of New York's history, although beekeeping was outlawed in 1999 in that city, when bees were added to a list of 'over one hundred wild animals, including hyenas, pit vipers and dingoes, considered too dangerous for urban life'.[30] In spite of the ban, a 'clandestine apiarist culture' survived until beekeeping was legalised again in 2010,[31] and is again flourishing, with hives in gardens and parks, on rooftops, balconies, and pretty much anywhere they can be located. Counter-intuitive as it may seem, New York provides a wide range of flowering plants for bees, and the different flowers available in various parts of the city produce distinctively regional honeys (South Bronx Honey, Manhattan East Village Honey, etc.). Even when 'housed' in a hive, bees are essentially wild creatures, and as one beekeeper pointed out, 'I think that "beekeeping" is something of a misnomer [. . .] You can't keep the bees. If they want to go, they're going to go.'[32] At the same time, the pollination of multiple plant species is dependent on bees, so their presence in New York draws sharply contrasting responses. As urban dwellers, some New York bees have at times found alternative sources of nectar in sweet factories, leading to unusual colouration and taste in their honey. '"We look at the city and think of it as being inhospitable to wildlife," said Kevin Matteson, an urban ecologist who has studied bees in New York City. "But when you look at it and think of what bees see and what they need – it is absolutely hospitable to wildlife." '[33]

There is a long history of farming vegetables in New York, often on vacant lots, including in Harlem, in which crops that had travelled on the slave ships, such as 'okra, black-eyed peas, watermelons, and certain beans and yams', were cultivated.[34] Growing vegetables became especially important during times of recession, and was actively promoted by the state during the Second World War, though it has continued both officially and unofficially ever since. With tragic irony, successful vegetable farming was often its own undoing, in that the conversion of vacant lots from dumping grounds to well-tended gardens frequently reduced crime and increased property values to the extent that the lot became desirable again for development. Intriguingly, experience from the inner city suggested that what was called 'rubbled soil', which contained old bricks, proved to be an 'ideal growing material' when combined with organic matter, apparently because of the lime, clay and sand in the bricks' composition.[35] Shulman remarks that '[b]y the 1980s, the most quintessentially urban city in America had sprouted as many as 1000 community gardens and 10 000 gardeners',[36] though urban farming of vegetables on vacant lots has always been vulnerable to the vagaries of city planners.

While few urban dwellers are likely to be offended by a vegetable garden, the concept of the 'farmed' as 'wild' is profoundly evident in responses to the raising and butchering of animals in the city. New York has a long history of meat production and slaughter. Shulman quotes the response of a European visitor as early as 1830 who was 'appalled that slaughterhouses were "scattered all over many populous districts of the city," filling the air with "the most noxious effluvia"'.[37] She points out that '[d]espite officials' best efforts, by 1842, roughly 10 000 stray pigs wandered city streets, and before a decade had passed, the number had doubled'; and as recently as the late 1950s, 'cows, pigs, and sheep were still walking through Midtown Manhattan en route to massive slaughterhouses'.[38] In very recent times, the 47-acre Queens County Farm Museum in New York, which had been a kind of petting zoo, began to raise animals for slaughter to the delight of those who prized ethically farmed local meat, and to the horror of those who insisted that killing animals had no place in the confines of a modern city.[39]

While the bulk of slaughtering and meat packing has now either been pushed to the very edges of the city, or far beyond its confines, so that its inhabitants experience meat only as a prepackaged commodity, Shulman's research alerted her to the fact that 'some of the city's eight local live poultry houses also dispatch goats, lambs, and sometimes cows'. These cater especially to the needs of immigrants who 'expect to look into a living animal's eyes, feel its flesh, and know it's fresh before killing it to eat', and she argues that large numbers of animals 'are still raised for food in backyards, basements, community gardens, and city streets'.[40] City slaughterhouses catering to halaal and kosher needs, despite much resistance, especially from upmarket residents, also continue to operate. Shulman concludes: 'If you go looking, you can find enough urban backwoodsmen raising meat in New York City to give you the sense that slaughter could lurk in any Queens estate or Bronx high-rise. Women and men in every borough still raise chickens, turkeys, ducks, and geese for meat.'[41]

New York is also a city of harvesters of wild fish. It is a city that was previously blessed with abundant marine resources, including superb oyster beds, but pollution has destroyed those. There are now signs on docks and piers warning: 'PREGNANT WOMEN, WOMEN OF CHILDBEARING AGE AND CHILDREN UNDER 15 YEARS OLD SHOULD NOT EAT FISH OR EELS CAUGHT IN THESE WATERS';[42] and yet many people continue to catch and consume fish, as a source of free food. They fish with rod and line, seine nets, crab traps, handlines, even in one account by hauling out car wrecks, which proved home to multiple forms of sea life that could be eaten.[43] Again, it is predominantly, but not exclusively, immigrant communities who are involved:

> All around the city's edges, Indians, Jamaicans, and Italians eat the smaller fish, the little shiners. Puerto Ricans and Bosnians and Poles go for bluefish and striped bass, the bigger the better. People will bring a black trash bag to the docks and toss in everything they catch. They will hitch a Styrofoam cooler to a luggage cart to pack with seafood. They will cozy their fish against an ice pack for a long subway ride through the boroughs. Most are not destitute, but they look to the waters to help them in their struggle to get by.[44]

One study suggested that in families where people fished, the family members ate an average of 9.5 fish per week, despite the fact that the state Department of Environmental Conservation recommends that women and children eat no fish from city waters at all, and that men eat them in restricted qualities calculated by species.[45] Those who consume such fish frequently point out that they have done so for many years, to no apparent ill effect, and in some cases so have their parents. Having to choose between the immediate relief of a full stomach and the long-term dangers of chemical poisoning and carcinogens is not something anyone, especially a parent, should ever be forced to do.

In this case, as in so many others, we have defiled the wild, and betrayed ourselves as a species.

8

Wild Fish

We tend to see trout as finished products, too, as if around
1880 or so, Evolution dropped them off at our door and
said, 'Have a nice time.'
— Paul Schullery, *Royal Coachman*

It is well to have some water in your neighbourhood, to
give buoyancy to and float the earth.
— Henry David Thoreau, *Walden*

The line felt its way through the sea like an extension of
my senses, an antenna attached to my skin, twitching and
trembling.
— George Monbiot, *Feral*

In this chapter I consider what it means to talk of 'wild fish', and for
reasons that I explain during the chapter, my main focus is on 'wild
trout'. I consider 'wildness' not just as a quality that is valued in such
fish, or in describing them, but also what 'wildness' means as a strategy
for 'managing' fish populations and the ecosystems that they inhabit.
My discussion is restricted to freshwater fish populations, as the sheer
scale of ocean environments means that 'wilding' in the sense of
proclaiming marine reserves and controlling fishing are really the only
'management' options. 'Restocking the ocean' would seem at best a
quixotic conception.[1]

'Wildness', whether as a mark of value or as biological process, is not something that one encounters readily at the frenetic speed at which most of us live our urban lives, so I begin the chapter with two accounts of taking time to seek out 'wild fish' in distinctly different locations – Somerset East and Rhodes. I then try to tease out some of the implications of these experiences in relation to arguments about 'wilding' or 'rewilding' waters.

* * *

A man scatters seed on the ground. Night and day, whether he sleeps or gets up, the seed sprouts and grows, though he does not know how. All by itself, the soil produces grain – first the stalk, then the head, then the full kernel in the head. — Mark 4: 26b–28 NIV

* * *

I left Cape Town under dire water restrictions, with gale force winds driving runaway fires. The smoke fogged out the sun, bathing everything in an ominous orange haze, and you could taste the ash on your tongue. Up the N1 to the Huguenot Tunnel, the wind pushed the bakkie around like a plaything. Through the tunnel, the Molenaars River was a thin stream, more rock than water. Further up, the Jan du Toit's River was dry, as was the Breede. The onward journey revealed one empty river bed after another, and dams either bone dry, or reduced to pitiful puddles. The river in Laingsburg, once responsible for such devastating floods, appeared to have been turned in part into a dirt road. The first water I saw was in the Leeu River at Leeu Gamka, ironically deep into the Great Karoo. Around Beaufort West, the landscape was more bare earth than Karoo scrub, and pied crows seemed to be the only things moving.

It was a strange journey to be undertaking in search of wild trout, but that was exactly what I was doing. The waterless landscape seemed to mock the fly rods on the back seat, and dry the saliva in the mouth.

The advertisement, 'Wild Fly Fishing in the Karoo', had caught my eye some time ago, as counter-intuitive as the desert salmon in the novel *Salmon Fishing in the Yemen*. This was my second pilgrimage to the trout of Somerset East, followed by a second visit to the 'wild trout' of Rhodes, a more obvious location for trout in the well-watered montane region, which is host to the annual Wild Trout Festival.

The trout of Somerset East are 'wild' in the sense that they are stocked as fry, fingerlings and, at the largest, yearlings (depending on the nature and predator/fish populations of the waters stocked), and then left to get on with it, growing in some cases to in excess of 5 kilograms even in the rivers.[2] The rate of attrition is high. About 5 per cent of the fry make it to adulthood. Fingerlings are stocked when the water contains large crabs or dragonfly nymphs, which would easily make a meal of the fry, and the survival rate is about 10 to 15 per cent. Yearlings are typically stocked into waters that contain large platanna frogs, for which a fingerling is a readily accessible meal. The food supply in the waters is abundant, and the trout thrive in this region.

In the tributary sources of the Little Fish River, there are also smallmouth yellowfish, but only in the lower sections, as man-made barriers and the geology of the river bed prevent them from moving up to the source. It is a strange landscape: much of the 'river bed' is dry, as the water travels underground until rocky outcrops force it to the surface, where it collects in big, deep pools (apparently scoured by dinosaur-like reptiles many, many years ago) only to disappear underground again until another rocky outcrop forces it to the surface to form another pool. These pools are stocked with trout. Travelling to the fishing waters, one is puzzled by the kilometres of dry river bed. When taking clients fishing, local guide and pioneer of fly fishing in the region, Alan Hobson, likes to stop on one of the bridges crossing the Little Fish River, where one sees thickets of acacia trees and stretches of sandy river bed, and announce nonchalantly, 'This is the river we'll be fishing for trout.' But the water is kept sufficiently cool on its underground journey that trout thrive in these pools even in the baking heat of the Karoo summer, where daytime temperatures are regularly over 40 degrees Celsius. There are some sections of the rivers in which the trout can breed, but the bulk of the fish grow from stocked juveniles. They are beautiful, well-conditioned fish, as spooky as any self-sustaining river trout populations I have encountered, and with a strength and tenacity deriving from their rich food sources to satisfy the desires of pretty much any angler after wild fish.

But the 'wild' experience is also in the landscape that these fish occupy. To the west of the town, you are in montane bushveld, rocky, scraggy, with vegetation that claws at your clothing as you pass. On a trip with Alan to fish these waters, we encountered kudu, nyala, reedbuck, black eagles, a martial eagle, blesbok, various hues of springbok including black, waterbuck, scimitar oryx, exotic red deer and fallow deer, buffalo, giraffe, spring hares and bushbuck, as well as a very large boomslang in the road in front of us, which we admired from the safety of the double cab, and which soon proved as wary of us as we were of it, and melted into the grass on the verge in one smooth glide.

And then there are the trout . . .

In a landscape that at best should be home to the odd mudfish or barbel, there are magnificent places in which to catch trout. At first your mind refuses to believe that there are salmonids of trophy proportions here, amid wag-'n-bietjie and haak-en-steek thorn bushes. But they are there, and thriving. And what awaits the stealthy angler is challenging fly fishing for trout in a landscape as quintessentially wild and South African as they come. And then, just as your mind feels as though it has been stretched beyond its capacity to absorb this phenomenon, Alan casually mentions that the top section of the river is not accessible to fly fishers at present 'because of the buffalo up there'.

Take the road out of Somerset East in an easterly direction, follow the road up towards the waterfall at which Walter Battiss, the abstract artist who painted exquisite watercolour landscapes, used to commune with the universe, and you are in another world. The steep mountain forest landscape is reminiscent of Hogsback, though not as wet. Alan and I followed what could only in the most generous estimation be called a 'road', deep into the forest. The area is a national heritage site, and I was entertained with a lively commentary about the various tree and plant species, some endemic to the region. It was a landscape of wonder, one in which the sacral is almost tangible.

Jolting to a halt, we changed into waders and boots, loaded up our gear and started the trek down to the river. Yellowwoods, sneezewoods, cabbage trees, wild pears, olives, all lined the path in wild profusion, some festooned with old man's beard and lichen, and there were several stops for discussions and pictures. We watched the black eagles soaring above the cliffs, and then almost grazing the cliff faces with their wing tips as they raced along them. On the path down to the first pool, Alan was concerned by the apparent clarity of the water, which meant the fish would be spooky, and also by the presence of cattle by the pool, which would surely put the fish down.

But when we arrived, there were fish rising everywhere, apparently tucking into the insects following the cattle, which seemed to be functioning as a mobile bovine insect hatch. After a hasty tackle-up I sent a tiny dry fly (a #20 Para RAB of Alan's tying) into the edge of the melee, and was met by a confident head and tail rise. A lively battle ensued before an exquisite rainbow of about 250 grams was netted and released. For reasons that I shall not try to justify, my casting was off-song, so I messed up the next few presentations. But the next decent one produced another lively take and skirmish, and another beautiful fish was released. Alan judged that these were the fish he had stocked about eighteen months ago as fingerlings. The cattle moved off, and so did the hatch, so I switched to a subsurface nymph, and hooked (and lost) a fish of about 10 centimetres, which seemed inexplicably not to have followed the growth trajectory of its pool mates.

With the surface action dying, we opted to hike further up towards the waterfall pool, which was our final destination. 'We'll just be hiking,' was Alan's instruction as we set off. That was until we approached a pool, barely a few feet deep, with two trout feeding actively but warily. After the loss of confidence from the fluffed casts on the previous pool, and frustrated by my attempts to get ready to cast in the tall grass that kept us concealed from the trout in the pool, I gave the rod to Alan in exasperation. 'You fish this one.' The two fish moved skittishly around the pool, seeking cover under an undercut rock and then darting out for a food recce. Alan pricked the larger one, and then a few casts later there was a good bend in the rod, as he hooked the smaller fish. The fight took place in crystal-clear water, and the second fish followed the hooked fish closely. After a quick photograph, it swam off to resume its activities. As we hiked past the head of the pool, we watched it rush for cover as it spotted us.

The walk upstream to the waterfall pool was a combination of rock hopping, wading and scrambling up boulders. Still in the final stages of recovery from a serious illness that had put me in hospital the previous year, I was a little puffed by the time we reached the pool. Breathtaking walls of red-yellow-grey rock rose on either side of the waterfall, which dropped a sparse but steady stream down 90 metres of cliff into the pool, splashing on its descent across agapanthus and watsonia plants rooted into cracks in the rock. The rock faces were all you could see, and the water seemed to fall out of the sky. It felt like a place of reckoning, a place where things began, or ended. Baboons uttered their territorial barks far above us, as we settled ourselves down for lunch and a drink. The sense of being in the presence of wildness and wonder was overwhelming, and my attempts to explain to Alan how it felt for me to be there seemed inadequate, even disrespectful. We ate our lunch mostly in silence, with occasional remarks about the thunderclouds ridging in above us, and the attendant flash and thunder roll.

Having come so far, we felt that despite the incoming storm we had to have a go at the fish, which had been rising sporadically throughout our lunch. We boulder-hopped to the top of the pool where, after some very careful low-profile manoeuvring, Alan put me into a fine casting position. Having been duly prepped about the importance of not false-casting over the fish, I fluffed the first cast. Cast number two sent the dry fly straight out along the bubble line, which was formed by water falling from above. I felt rather than saw the take, so delicate was the sip, and I was fast into a strong fish. In the clear water I could see that this was a fish I did not want to lose. When it finally came to the net we judged it to be in the region of a kilogram. It had the exquisite fin colouration indicative of some cut-throat genes in its lineage, with a broad, powerful tail, which would shame the diminutive appendage of any hatchery-grown fish. As trout go, it was one of the most special and beautiful I have encountered, easily matching in the wariness of its pursuit any of the wild-spawned trout I have caught in the Mooi, Umgeni or Umzimkulu rivers.

The weather was increasingly ominous overhead and as the road out was treacherous when wet we decided to call it a day, hastily disassembling rods and heading back. We were soon drenched with sweat – a combination of pre-storm humidity and a desire to be on the road before the storm broke, which increased our pace. Once we had climbed out of the river gorge, the stone underfoot gave way to sand, and Alan pointed out the fresh spoor of a leopard, overlaying the prints we had made on our way in.

We stopped at the farmer's house to fill in our catch return and pay the gate fee. After a conversation about the fish, trees and plants that we had encountered in the waterfall gorge, he took us to see the fine specimen of a Cape chestnut tree in full flower in his garden, the only survivor of many more that he had cultivated from seed. As we left, fat drops of rain began to fall. Having lived through months and months of drought, I relished the sensation of water on my face. I was reminded of a conversation with a friend of hers that a colleague, Jane Taylor, had reported to me shortly before this trip: that the sky had forgotten how to rain, like an old man trying to piss.

* * *

Waves of student protest have wracked South African universities, with vice chancellors assaulted or threatened, buildings torched, and students and staff intimidated. My stomach has contracted into a solid knot, as all we have worked for over so many years, all that we have built, is threatened by mindless violence. I need some respite, a retreat to a place where the rocks, fish and kelp are as they always have been, and will be long after I am gone, where the wind tugs spray from the breaking waves, and oystercatchers probe the mussel beds with fine red beaks.

*   *   *

Above the waterfall pool, and following a set of roadways that seem to take you ever upwards to the top of the world, is Mountain Dam. I have fished the Highmoor dams in KwaZulu-Natal, which at that time seemed to me the pinnacle of altitude, but Mountain Dam tops that. On the road up, you can see across the plains below, and look clear across the Battiss waterfall from above, but you still have many more metres to climb to the dam. Here you are faced with mountain grasslands, reminiscent of areas around Underberg or Swartberg in KwaZulu-Natal. The dam is approached through an inexplicably complex set of farm roads and gates, across the land of several different owners, but when you arrive, you find a stillwater of the most extraordinary clarity, fed by mountain springs seeping out of the mountain top, and inhabited by trout that are magnificent in terms of condition, strength and appearance. It must be one of the loveliest fisheries in South Africa.

On my first trip, we opted to fish small nymphs across a shelving bank and weedbed, with the wind blowing the insects towards and across us. I landed a fish of about 800 grams. It was beautiful, with lively markings and the broad tail and fin structure that speak of a life spent chasing prey under conditions of wind and current (yes, despite their names, stillwaters have currents). The same can be said of the fish at Bluegum Dam at Lourensford, where it would sometimes take a geneticist to tell a wild-spawned rainbow from one stocked at small size and now apparently living a life (and achieving a biomass and biomorph) equivalent to its wild cohabitants.

On this occasion, we tried a few unsuccessful casts into the deep water off the wall, before moving across to the promising-looking shallows, shelving down to weedbeds, which just had to hold fish. We took a couple of lively fish, which seemed to have a predilection for

Alan's wonderful beetle flies, perfectly mimicking a water beetle that we had spotted earlier. The fishing was immensely satisfying, requiring some persistence, but with admirable rewards. Releasing my final fish of the day, I looked down at a creature that had grown fit and strong on the abundant insect life of the dam, clearly had the wits to evade the omnipresent cormorants and much more reticent otters, and which sported a tail that would have done a fish twice its size proud.

On the drive out, the red-footed kestrels were out in numbers, dropping from the sky to feast on the brilliantly red and blue locusts that seemed to be everywhere. As we approached the last of the gates on the farm road, I spotted a familiar shape through the grass stems, but my brain seemed reluctant to process what I had seen. Then, in confirmation, a black eagle heaved itself into the air, wings clipping the grass stems as it disappeared over the ridge.

Those fish, those environments, were wild enough for me, and I suspect for just about anyone, in the experience of catching fish that had grown up in healthy ecosystems, being left to fend for themselves against predators and drought, and being part of a montane system of extraordinary biodiversity. I left Somerset East utterly satisfied, and not a little humbled, I might add.

* * *

He was with the wild animals, and angels attended him. — Mark 1: 13b NIV

* * *

The wild fish at Rhodes are a different story, though. My first experience of pursuing trout at the Rhodes Wild Trout Festival in 2016 was bitter-sweet. After a fascinating but largely fishless sojourn at Somerset East, I arrived at Rhodes in teeming rain, having managed to put my bakkie into a ditch on the treacherous clay road into the village, despite driving at the speed of an arthritic tortoise (miraculously the only damage was a scratch to one of the rims, and the diff lock got me out).

In describing its waters, the Wild Trout Association says that 'the Eastern Cape Highlands, technically speaking, the Real Southern Drakensberg, is the true domain of the wild trout of southern Africa'.[3] Advertising hyperbole aside, there is some substance to the claim. The philosophy and management practice of the Wild Trout Association are simple. This is an area, encompassing Rhodes and Barkly East, which has approximately 350 kilometres of rivers with self-sustaining populations of trout and some yellowfish (the first trout in the Barkly East area were stocked as early as 1910, and the remainder of the area in the mid-1920s),[4] and the Wild Trout Association simply manages access to the rivers through a booking and beat system, involving partnerships with riparian owners, so limiting poaching and letting the rivers and their fish 'get on with things'. It is a 'wilding' or 'rewilding' project on a grand scale. There is no stocking of fish, trimming of bankside vegetation, making paths, or the like, and the rivers are simply allowed to function as 'wild' ecosystems.

In the days to follow, I met some fine people, and fished some exquisite-looking stretches of river. But to say that the fish were few and far between would have been overly kind. Clearly the drought and extreme heat of the previous two years had affected the fish population badly, and little if any spawning had taken place. We worked our various allotted stretches of river hard for the four days of the festival and ended up with the following tally among our group of four anglers: Andrew Mather: 1; me: 1; Terry Andrews: 0; Quentin Austin: 5 (at more than one fish a day we regarded him as something of a cormorant). The water was in exquisite condition, after a brief period of discoloration from the rain, and the landscape is one of the most beautiful I have ever encountered. But teeming with fish the rivers were not.

In what seemed to be an appropriately metaphoric ending, we concluded our final day at the very top of the Glen Nesbitt beat, in which Terry spotted a beautiful fish of about 40 centimetres holding about 3 metres down at the very bottom of a deep pool, the very last piece of water before the beat ended with an emphatically sturdy fence. With Terry guiding his cast, Quentin tried unsuccessfully to catch that fish. Andrew and I had had about enough by that stage, and were also sufficiently certain of the outcome to flop down in the grass instead for some rest. The fish was utterly unmoved by our efforts, and simply kept station as before.

Eventually we gave up, and wound our weary way back to the truck. Suitably chastened by failure, we comforted ourselves with the fact that the uncatchable fish would be a fine contributor to the spawning season to come, and hence the repopulation of the river. At the closing dinner that evening, one of the organisers joked that the fishing had been so poor because someone in the club had absconded with the money set aside to stock the river. It was something of a wry joke to the largely fishless congregation of anglers gathered there that night, but also a salutary reminder that droughts and periodic population declines are integral to the functioning of ecosystems, and actually serve to toughen up the gene pool. So if you want 'wild fish', you will have your share of fishless days.

My second visit to the Rhodes Wild Trout Festival, in 2017, was rather different. Good rains and snowfall in the previous year had led to a much more successful spawning season, and the fish were more plentiful, though far off the numbers reported from previous years. Andrew Mather and I paired up with Sheena Carnie of *Flyfishing* magazine, and we managed a fair number of small fish in the days to come, many on the dry fly. They were beautiful fish, iridescent in colouration, and a joy to behold. On our second day, Sheena was not well, but friend and festival guide Peter Brigg joined us. Watching Peter fish was an utter revelation for Andrew and me. As the festival progressed, the water levels in the rivers dropped, making the fishing increasingly challenging. Those fish would see us coming from far off, and only the stealthiest wading and casting paid dividends. On the drive to our allocated beat one morning, we came across a relatively recently killed rabbit, evidently struck by a passing car, which Andrew insisted on picking up, rabbit fur being highly prized by fly tiers. (I say 'relatively recently killed', as the aroma of dead rabbit gently pervaded the Subaru, and reaching into the boot for a piece of equipment meant being confronted by the macabre sight of an extremely rigor mortised rabbit's leg protruding from the woefully inadequate Checkers packet into which it had been shoved.)

Towards the end of the festival, the three of us were offered the Ben Laws beat on the Upper Bell, again with small-stream maestro Peter Brigg. This was to be an exercise of pursuing wild trout par excellence. At this point, the Bell is a tiny stream, not more than 30 centimetres or so deep in places, in high, open grassland. Close to the river, the vegetation gives way to large raised tussocks of steekgras, with unexpected holes and marshy sections between them, which makes walking a perilous exercise. When we approached the stream, all but our very experienced small-stream guide were a little flummoxed. The river was so low, clear and narrow that you would have sworn not even a tadpole would survive there unnoticed.

But we obeyed Pete's instructions and got fishing. As I entered the tail of a pool not much more than 10 centimetres deep, aiming for the small gurgle of water at the head of the pool in which I thought a fish might, against all probability, lie, a trout of about 30 centimetres broke cover from under the shade of the bank to my right in water barely deep enough to cover its dorsal fin, and shot off towards the top end of the pool. I realised then that we were into a completely different game, with quarry so attuned to the finest movement or sound out of the ordinary that they had bolted long before we even imagined where they might be. I felt clumsy and inadequate in this environment of infinite subtlety and delicacy.

Up a series of thin pockets of water, I found one slightly deeper, with some moving water disturbing the clarity. To my utter surprise, my first cast produced a rise, a small fish splashing at the dry fly – but a fish nevertheless. A couple of casts later and a brief on-off after a rise from a better fish convinced me that this might not be utterly impossible. Then I reached a pool containing about ten fish, including some good ones. It was so clear that the fish could have been swimming in air, which they might as well have been for all my attempts to catch them. Even the most delicate cast had them all streaking for cover under the rock by the tail of the pool from which I was casting. After a minute or so, a small fish would reappear, and conduct a frenetic recce. Then three slightly larger fish would emerge, and set up station at 45 degrees to the current just about 4 metres above me. Even the gentlest cast would send them all shooting back for cover, and then I would have to wait for the game to begin again, first with the lone fish on its recce . . . My cunning plan of casting upstream and waiting for them to come out and find my fly was flummoxed by the fact that, even with so slow a current, it was impossible to keep the fly in their zone for as long as it took for them to reappear. Worrying about my sanity, in what seemed like an infernal game that I was destined always to lose, I gave those fish best and set off upstream.

Pete put me onto a beautiful-looking pool with a small waterfall, which screamed fish. Dropping my fly in some slack water before stripping off some line to cast, I was surprised by the insistent tugging of a decent fish that had engulfed my fly and set off at high speed. I lost it after a few seconds. Further casts proved fruitless, so I gave the pool to Pete, who promptly landed a plump and lively fish on the same fly as I had been using. The best I could offer was the loan of my net and to hold the fish for a quick photograph session.

I continued my fishless journey upstream.

Andrew caught a decent fish after some careful stalking, and Pete winkled a couple out from the shadows of the undercut banks. We called a halt at a pool with a large flat rock shelf on which to sit, and ate our sandwiches at a leisurely pace, surrounded by high grasslands, enjoying the warmth of the rock and the coolness of the water after the efforts of the morning. Our thoughts were lazy and full, and we were in no hurry to move. We dangled our legs, wading boots and all, into the water, while small trout milled around the bottom of the pool – darting, squiggling arrows – and a bigger fish unexpectedly shot across the pool to seek cover under a rock.

Eventually we headed back downstream, with the determination to fish those spots that had looked likely on the way up. I caught a small trout at the waterfall pool at which I had lost one and Pete had caught one. The rest of the party eventually congregated at the car, weary after the day's exertion. Andrew had been collecting sheep's wool and horse hair from the barbed wire fence for fly tying, apparently having become a little feral after the rabbit episode. Pete finally arrived, with tales of having landed two superb fish from a pool in which he had lost a fish earlier in the day. He had such a glow of pleasure that I still wonder whether he hadn't met a mermaid instead.

* * *

The water was a little coloured, not muddy, but less clear than during the brilliance of its summer flow, and brought with it fallen leaves and twigs and dead fir needles. Most of the leaves twisted and swam and swirled a few inches below the surface – alder leaves, some black and rightly fallen, others still green, torn from the trees by winds that brought the fall rains; maple leaves, sodden dark brown and fast breaking up; willow leaves, long and slender, some yellow, some black. Under the leaves, deeper in the water, were the salmon. — Roderick Haig-Brown, Return to the River

* * *

My purpose in narrating these rather different, but equally satisfying, fishing experiences is to reflect on what it means to talk of 'wild fish'. I am partly concerned with the experience of pursuing fish that are, or are perceived to be, wild; but, perhaps more importantly, I am concerned with what 'wild fish' mean in terms of the health of ecosystems, the species they sustain, including the fish themselves, and the larger question of human presence in such landscapes.

In this respect, trout are especially interesting to think with, for the question of their assumed wildness or otherwise is fundamentally tied up with their identity as trout. If one of the most significant measures of wildness is existence in self-sustaining populations independent of human intervention, we could surely talk of wild bass, barbel, carp or yellowfish? But no one does. It appears to be only trout that are described as such, though it is true that the term 'wild' has been introduced more recently in relation to marine fish to suggest their sustainable status under SASSI ('Wild caught from sustainable stocks', reads the packaging on Woolworths hake, for example).

In *Are Trout South African? Stories of Fish, People and Places*, I discussed the categorisation of trout by fly fishers in terms of their wildness, from wild-spawned river fish at the top of the value chain, to pellet-fed fish stocked into put-and-take waters at the bottom. In this chapter, I am trying to shift the argument away from wildness as a mark of value or attribution (is this trout wild enough to be valued as a catch?), to an understanding of wildness as a management model, as fundamental to the biological processes of the ecosystems. Along with Monbiot, I am suggesting that we achieve the best results, all round, when we allow wild processes to operate in ways that are as far as possible self-willed or self-regulatory. Of course, wildness as value and wildness as operating principle may sometimes seem like the same thing, but they are not always the same, and the distinction can be crucial. (As Paul Schullery argues, when confronted with the full implications of what it means to prize 'wild' fish, many anglers become hesitant about how wild they want their fish to be, and are quick to call for remedial human action.)[5] In addition, following Monbiot, I would stress that this understanding of 'wildness' does not simply equate with 'naturalness', in that the ecosystems and the species they contain may look rather different from what was there at other historical periods: it is very specifically not a 'restoration' argument, but a 'rewilding' one.

Even human-created habitats can be allowed to function on rewilding principles. As we are not blessed in South Africa with many natural lakes, almost all dams, or what fly fishers call 'stillwaters' or in some cases 'lakes', are by definition artificial. They generally involve excavation, the construction of a holding wall, and either the damming of a river or spring, or location in a run-off area from which they catch rain water. But they soon become ecosystems in their own rights, with the growth of water weeds, the attendant development of insect and amphibian life, the attraction of bird life, frequently the introduction of fish species (sometimes, but not always, by humans), and, if we are fortunate, the appearance of charismatic species such as otters and fish eagles, as well as water monitors and various other species that love such aquatic environments. The point of dam construction is not necessarily to provide fishing, as dams are generally built to supply water for human consumption or for irrigation and livestock, but where they are managed as fishing waters, the aim is generally to keep them 'as natural as possible'. Sometimes there are misguided management efforts to control water weed growth by the introduction of grass carp or chemicals, which backfire spectacularly. Dams that are more lightly managed, in which, for example, weed growth is allowed in the shallow sections from which the water enters (assuming the typical triangular shape of most South African dams), usually support healthier ecosystems and hence provide better fishing, both because of the resulting increase in insect life in the weed and also because the weedbeds filter the water entering the dam, thus improving the water clarity. Intensely managed dams, in which fish are stocked beyond the carrying capacity of the water and then sustained by supplementary feeding, are subject to multiple and continuing problems requiring ever-more drastic 'solutions'.

* * *

The water is crystal clear and cold, and I can just see the weeds on the bottom, many metres below my float tube. The last two days have been fishless, and I am beginning to believe that the entire population has headed up the feeder stream to spawn. The surrounding mountains, still spotted with snow from the last fall, seem to hold the lake in the palms of their hands. Despite the absence of fish, I am at peace with the world, drifting across this water as smooth as glass, with only the dabchicks for company and the call of the resident fish eagle occasionally piercing the wintry silence. Up above the fold where the feeder stream runs, three pale shapes are visible as they cross a firebreak. Eland. And at that moment the line draws tight, and I am fast into a good fish. Seconds later, the line slackens, and both fish and eland are gone. It is what Mr Wordsworth called a 'spot of time'.

*   *   *

In his book, *Royal Coachman: Adventures in the Fly Fisher's World*, renowned US conservationist and fly-fishing author, Paul Schullery, offers some challenging and far-reaching insights into understandings of wildness, not least those held by fishermen focused solely on the successful pursuit of their chosen quarry. His loyalty to the healthy biological process rather than fly fishing at all costs is clearly stated at the outset:

> Under the National Park Service's policies and legislative mandates, the introduction of a large, nonnative predator like a trout into a small alpine lake is only wild in the sense of being wildly inappropriate. To an ecologist, the lake is not, in the fisherman's language, 'barren'. It is full of life, a unique and precious little ecosystem that has spent thousands of years developing its community of native species of invertebrates, reptiles, amphibians, and plants.[6]

Turning his attention then to what may be properly construed as 'wild', rather than 'wildly inappropriate', he begins with the conceptions of 'wild' put forward by the fisheries biologist Albert S. Hazzard in 1944 in an article in *Outdoor America* about trout fishing in national parks:

> To him, a trout that grew up in the water, even if it was planted there as a fingerling, was wild. That view was easily justified, at least for his purposes, because the fish had to become a functioning part of its new home in order to survive and grow; it had to find food and shelter and evade predators, thus becoming thoroughly adapted to the place. In Hazzard's view, any fish that achieved all that deserved to be called wild. On the

other hand, fish that were just dumped out of a truck tank into the water last week as adults, and were still swimming around in confusion when they were hooked, did not.[7]

That sense of wild, of having had to 'become a functioning part of its new home in order to survive and grow [. . .] to find food and shelter and evade predators, thus becoming thoroughly adapted to the place', would match the assumptions of many fisheries in South Africa (in the longer view, even of the very existence of trout in South Africa), including the waters managed by Alan Hobson in his 'Wild Trout in the Karoo' venture.

But Schullery points to the shifting goalposts in what we mean by 'wild', in that in his experience, 'Today, relatively few fishermen and even fewer managers would call a stocked fingerling wild, even if it spent several years in the water before being caught.' Our 'understanding of wild ecosystems has evolved', he maintains, 'so that we have higher expectations when we insist on wild trout in our streams'.[8] In this respect, there is now a shift from measuring wildness in terms of the time not that the individual fish has spent in the water, but how many generations of fish have lived in the water: 'A fish stocked as an egg or fry isn't likely to be called a wild fish; a fish whose grandparents were stocked into the stream is'[9] (the fish in the rivers of Rhodes and Barkly East have a lineage of around a century, so would properly be 'wild' in this definition). That shift is crucial, as it emphasises not the value of an individual fish itself, but the sustainability of the biological processes that have led to the hatching and growth of the fish population as a whole.

To complicate matters further, Schullery notes an additional shift in thinking among those involved with 'wild' fish, from 'wild' to 'native', and an emphasis at fisheries management conferences and in conservation circles on 'catching fish in their native settings'[10] (which is, interestingly, what Dean Impson told me, while I was conducting research for the book *Are Trout South African?*, gave him the greatest satisfaction). This is a major part of the appeal of fly fishing for indigenous species such as yellowfish or tigerfish in South Africa; you are fishing for species that have adapted to those specific environments over thousands of years.

Schullery insists that the drive to stock angling species where they did not previously occur has in many cases had disastrous effects on the ecosystems, which it is assumed will then support the non-native species introduced:

> As we have introduced non-native fish into fishless waters, but also in waters containing native fish, we have lowered an ecological eggbeater into some glorious native ecosystems, resulting in changes that, though they may have been wonderful for fishermen, were disastrous for these beautiful little worlds that have been cranking along without our help since the last ice age.[11]

He points out that with more recent emphases on biodiversity and the conservation of native ecosystems, 'wild trout fishing has become more and more a setting-dependent sport'.[12] But Schullery notes that even in such a view, 'Letting trout be totally wild, and enjoying them on the terms that "totally wild" implies, isn't as simple as it sounds.'[13] Herein lies the rub for many who claim to want wild fish, but may then decide perhaps they do not really want them 'as wild as that' when the full implications of 'letting the fish be wild' become evident. Schullery makes the assertion – blindingly obvious when one thinks about it, but seismic in its implications – that 'it is a common human vanity to assume that the earth as we see it today is somehow a finished product', and points out that in contrast '[a]ll the processes that shaped the American landscape, whether geological, biological or climatic, are still acting today, unless we change them or stop them, which we often do for good reason. Nature isn't done with our trout streams.'[14] And as the first epigraph to this chapter suggests, trout 'exist themselves in a state of change'. So, '[w]hen we make trout and trout streams into something we find more suitable to our tastes, we may be producing wonderful fishing, but we are not taking wild trout fishing to its logical conclusion'. The crucial question becomes 'how much dare we compromise the native ecosystem's integrity without ruining the wildness of our trout fishing?'[15]

Wild environments include droughts, floods, fires and temperatures that may be by turns scorching or arctic: all factors that militate against received assumptions about 'good' and 'consistent' fishing – vide my first trip to Somerset East and Rhodes. But these factors are fundamental to shaping the qualities of the quarry we value, as Schullery points out: 'The very forces that had so much to do with creating the trout we admire – the violent extremes of environment that provided these species with the tests that turned them into our favorite fish – are a critical part of their wildness.'[16]

He refers to the huge fires that raged through Yellowstone Park in 1988, and the diametrically opposed reactions to them within, for example, the conservation and fishing/tourist communities, pointing out that '[t]he same forces that brought us the fires of 1998 brought us wild trout. If we want authentic wild trout in an authentic wild setting, the fires are part of the price that must be paid.'[17] That conception of wildness requires that as fly fishers, and conservationists, hikers, hunters or anyone else who values such places, we readjust our time frames, from the short years of a human life to the centuries and millennia of biological and geological processes. As Schullery argues, 'Our sense of scale as fishermen is pretty short. We become impatient with nature, which has no sympathy for our short lives.'[18]

That is a profoundly humbling position, and one that radically decentres the human. As Gary Snyder states, analogously, in the forest it takes about the same time as a tree took to grow for it to decompose completely, and if 'societies could learn to live at such a pace there would be no shortages, no extinctions. There would be clear streams, and the salmon would always return to spawn.'[19] That approach can mean that, as I did, you may sometimes travel nearly 2 500 kilometres not to catch many fish, if you are committed to 'wild trout' and their 'wild ecosystems'; that you understand that the fish that survive drought or flood will produce offspring better suited to continued survival in that particular river, and will lead in the longer term to healthier fish populations. That longer term may, of course, be longer than our own lifespans, which is a scale difficult to get one's head around sometimes, but is really the only way to understand the health of the planet and its ecosystems.

Schullery's commitment to wild, native ecosystems is an extreme of the wildness argument, which he acknowledges is difficult or impossible to sustain, given the 'ecological eggbeater' we have dropped into the world in which we live. He, himself, does not pursue 'wild and native' as the only standard, acknowledging that there is value in fisheries that do not meet such exacting standards:

[I]n many places, for social, political, and ecological reasons, we have created trout fisheries that can never be 'wild' in any grand sense. Public recreational needs demand such fisheries, and they are often very good. I'm not preaching revolution against all other kinds of trout fishing here; not quite, at least. I'm just trying to clarify where we are in our search for wild trout.'[20]

And he also acknowledges that human intervention, in the form of stocking, has occasionally resulted in the survival of 'little gene banks of trout strains' in remote places, in contrast to other populations that have succumbed to the 'swirl of genetic confusion'. For example, 'Yellowstone's nonnative fish may in some cases be the purest examples of what those fish were once like.'[21] In South Africa, the Cata River and the Upper Umgeni are regarded as holding some of the oldest and 'purest' strains of Loch Leven brown trout, for similar reasons.

Schullery emphasises deep understanding of fish populations within the ecosystems that contain them, and a focus on their long-term health, rather than the short-term goal of providing 'good fishing' by the inappropriate stocking of angling fish, whether in terms of number, size or species. It is, at base, a 'wilding' or 'rewilding' argument, which finds local confirmation in the management strategies applied by Wolf Avni to the fishery and hatchery at Giant's Cup, just outside Underberg.

Avni became involved with Giant's Cup for the first time in 1987, but only moved there in 1989. The property centres on a large dam, fed by the Umzimkulwana River, a tributary of the larger and better-known Umzimkulu, with a self-sustaining population of rainbow trout, and a smaller holding of browns. There is also a trout hatchery that produces both live fish for stocking and table fish. When Avni first arrived at Giant's Cup, there was a house, a rondawel, a very basic fishing cottage, and a small hatchery, some of the structures of which can still be seen where the Umzimkulwana emerges at the bottom of the dam wall.

The original hatchery had been run by Teddy Morris, apparently more as a hobby than a serious business. Avni has increased the number of hatchery ponds, enlarged the existing ones, and improved waste and filtration systems. The water that returns to the Umzimkulwana below the hatchery is as clean as that entering it. He has also built several additional cottages.

For anglers wanting to pursue 'wild trout', Giant's Cup is one of the top destinations in South Africa. The Umzimkulwana was initially stocked between 1911 and 1922, and the fish in the dam migrate up the inlet into the river during spawning time. Avni says that in contexts like this, probably the best management strategy is to keep your hands in your pockets, and simply let the system get on with it.

As he points out, the fact that the trout are able to spawn naturally makes winter fishing difficult, as fish may be absent from the dam for some time. This is in contrast to dams that do not have feeder streams, as trout attempting in vain to spawn often congregate in the shallows and make quite easy targets for fly fishers (especially the cock fish become very aggressive, and will attack brightly coloured flies).

Those visiting Giant's Cup in winter and simply wanting to catch fish might wish that the natural population of the lake was supplemented with stocked fish, so that at least there was some reliable fishing all year round. That philosophy governs the management of many waters in South Africa and elsewhere, where the demands or wishes of users are paramount, and the environment is manipulated to satisfy these. (Even in the famous English chalk streams in the UK, hatchery fish are now stocked into some sections because of the demand for fish to be kept, and also for fish of a decent size.) In an interview, Avni said to me that his management philosophy is in direct contrast to that which characterises many other waters in South Africa. For example, what he describes as the 'self-absorption' of those who want endless supplies of large fish, far beyond the carrying capacity of the waters, has led to some extremely questionable environmental practices. He estimates the carrying capacity of the dam at Giant's Cup to be around 15 kilograms per hectare, which is about right for a high-altitude oligotrophic lake.[22] Models of trout populations suggest that there will be at least two 5-kilogram fish swimming around in the dam at any time, which is enough to get the attention of most fly fishers. In stark contrast, the well-known fish scientist Bob Crass refers to a syndicate in which, with artificial feeding and intensive stocking policies, anglers were allowed to keep 100 trout per weekend (I have no idea what you would do with all of those), and the fish yield per hectare was 500 kilograms![23]

At a place like Giant's Cup, or the streams of the Rhodes/Barkly East area, or any of the other waters in which there are breeding populations of fish not supplemented by stocking, the philosophy is the converse. The processes of the ecosystem are paramount, and human expectations must recognise these. The same could be said about waters in which small trout are stocked within the carrying capacity of their ecosystems, and left to grow on as the system allows.

And a properly working ecosystem is a beautiful thing.

At base, one is asking the question, for whom do ecosystems exist? As I have argued earlier in this book, saying that the benefit of humans is not an issue is nonsensical, and actually has no scientific logic. Human life itself is dependent on the existence of healthy and sustainable ecosystems, as we are learning to our peril. But the assumption that environments or ecosystems can or should be made to serve human interests or wishes, regardless of the effects, is a recipe for disaster, as two centuries of rapacious destruction in the name of human progress, pleasure and greed have demonstrated.

If you have a properly functioning ecosystem that supports a population of angling fish species, and you introduce strict catch and release practices, you have a resource that requires little or no management or financial input, and is readily exploitable by local communities in generating income, provided, of course, they have access to the riparian rights. In several places across the globe, local communities have realised, often with the help of conservationists, that the worth of a live fish that can be caught and released a number of times by a sport fisherman is many times that of a dead fish sold at a local market, so that responsible environmental management that preserves fish stocks and their ecosystems is understood to be in the economic interests of people who inhabit them or their regions. Green and brown agendas converge, in such instances. (I have referred to two such projects in South Africa, involving trout, but there is the potential for many more, including in estuarine areas.) There are, however, also many examples where such arguments have been used by upmarket fishing companies to secure sole access to regions, which brings no benefit whatsoever to local communities, who end up being deprived of both local fishing rights and participation in the tourist industry.

*   *   *

Student protests have brought waves of violence and destruction to my university campus. Buildings have been stoned, fires set, and fear lurks in the gut. With the threat of new outbursts, the security staff has locked down my building. I do not want to be caught inside during the next wave of violence. I feel vulnerable, trapped and at risk, as I find locked door after locked door blocking my exit. Something in me shifts. I am a creature seeking a way out, and I now look not for doors but apertures. At the end of the hall is a smashed panel in the bottom half of a glass door, and I wriggle and writhe my way through jagged shards into the open.

*   *   *

The logo on the side of Avni's bakkie reads 'Another load of wild spawned trout'. Marketing catchphrases aside, the statement raises the interesting question of whether it is possible, paradoxically, to 'breed for wildness'. It is common knowledge that trout breeding has been manipulated for the demands of the table, so that fish become increasingly shaped like a rugby ball rather than a torpedo, producing larger fillets and a bigger proportion of edible body mass per fish. Is it possible to breed for the contrary, for 'wilder' fish?

It is certainly possible, as the table fish example shows, to breed fish for certain physical characteristics, including those that may make for better sport fish. Martin Davies of Rhodes University's Department of Ichthyology has for the last three decades been pioneering innovative ways of breeding trout, not only by using porta-pools in an urban setting, but in terms of developing certain desired qualities in the fish. In part, by introducing cut-throat trout genes into the mix, Davies has produced – judging by the fish I caught in Somerset East that began life in his hatcheries – strong, torpedo-shaped fish with big, wide tails and exquisite fin colouration.

Avni's approach to this matter is extremely simple, in that every fish produced in his hatchery has at least 50 per cent wild genes. He traps hen and cock fish as they exit the dam to spawn up the Umzimkulwana, and then ensures that wild eggs are fertilised with hatchery milt or vice versa. This approach certainly provides a solution to the problem of the lack of genetic variety that I discussed in relation to salmon hatcheries and attempts to restock rivers in North America, though the other shortcomings in those approaches remain seemingly insurmountable. Avni observes that when you hold a wild fish to strip it for the hatchery, your hands brace against the basal muscle structure, and you can feel the strength of the fin tissue. In hatchery fish, that basal muscle structure is not evident, nor is the fin tissue strong, he says, so your hands tend to slip back off the tail of the fish.

But what about the instinct for self-preservation, for lack of which hatchery fish are frequently (and sometimes unjustly, in my opinion) maligned? Tom Sutcliffe recounts the experience of visiting the hatchery at Kamberg, when it was still breeding trout many years ago.

On approaching two different ponds containing small brown trout, he noticed that the occupants of one seemed to vanish into the shadows at the first sign of human approach, even from many metres away, while those of the other were clearly alarmed by human presence, but seemed to mill around in confusion in fairly plain sight. On enquiring from Rob Karssing, who was running the hatchery, what the difference between the two ponds was, he was informed that the first contained fish that had recently been bred from fish caught in the headwaters of a tiny stream, while the second were the offspring of brood stock that had originally been caught from the Mooi River some years ago. The second lot appeared to have 'gone tame'.[24]

And then, just when we think we have gotten a handle on this subject, something like this happens.

My son, Mike, and I decide we want a change of scene for our fishing this weekend, so we head off to the little dams at Eikendal, which we hear have been fishing well. A quick stop at the fly shop to buy our rod tickets has us in conversation with Philip Meyer about the fishing. 'Try the bottom dam,' he suggests. 'The fish are feeding actively. Use a small nymph suspended under a dry fly.' As we approach the dam, the fish are feeding hard in the shallows. There are boiling rises and porpoising fish everywhere, in water barely 30 centimetres deep. Two hours later and we haven't touched a fish, despite having worked through the contents of several fly boxes. We know that these fish were stocked last week from Gerhard Compion's hatchery at Lourensford, but they are keyed into a hatch of microscopic insects with the finesse of fish that have never seen a hatchery pellet. Finally, a fellow fly fisher gets a take. 'What are you using?' we ask. 'A size 22 caddis,' he says (about half the size of your pinkie nail). We return to the shop to tell Philip that his fish have suffered no damage from our efforts, and comment on their selectivity. 'Yes,' he comments, 'Gerhard seems to breed wildness into his fish.'

Shifting attention from genetics and breeding, can fish, and animals, 'learn' wild behaviour? Among many others, Sutcliffe has pointed out that trout that have seldom, if ever, seen a fly fisher can be ridiculously easy to catch in their apparent naivety; others have written of newly 'discovered' bonefish flats offering cast-for-cast fish. It is certainly the case that on heavily fished waters, fish become much more skittish and selective in what flies, lures or bait they will take. In some cases, there have even been reports of populations of stillwater trout becoming almost entirely nocturnal under heavy fishing pressure (regardless of whether they were stocked or 'wild spawned'). An obvious comparison would be the fact that animals in reserves or sections of reserves, in which no hunting is allowed, exhibit little or no alarm at the presence of people or vehicles, whereas those in regions that are hunted flee at any sign of either. Conversely, if human presence is associated with food – carrots in the Karoo; pellets in the trout pond – animals can swiftly be 'dewilded' (which is not the same as tame, as I pointed out in a previous chapter).

Wild or not, what is not at issue is the plasticity of trout genomes, which has been extensively discussed by among others, Danie Brink, of the University of Stellenbosch. As Avni put it to me very simply, 'They will evolve to survive.' Negatively, this has issued in the kind of eggbeater genetic whirl referred to by Schullery above, and described in detail by Anders Halverson in relation to rainbow trout in the book *An Entirely Synthetic Fish*. Positively, the 'same' genetic make-up of the species *Oncorhynchus mykiss* (broadly 'rainbow trout') has led to some very different outcomes. On one end of the scale, we have the development of the highly coloured and diminutive golden trout found in Golden Trout Creek, Volcano Creek and the South Fork Kern River, averaging 15 to 30 centimetres, with a bright colouration that is perfect for the shallow, pebbled streams in which they live (they are the state fish of California). On the other end, there is a strain of much larger green-shouldered rainbows of British Columbia that grow to massive proportions on a diet of minnows, which they hunt in shoals. Even where they have not evolved into subspecies, the very different life cycles, physical appearance and diet of populations of rainbow trout across the globe, or even across South Africa, testify to their extraordinary versatility and adaptability. If the ability to survive is a marker of wildness, trout seem to have wildness in their genes.

9

Wild Ethics

What we find in a soulmate is not something wild to tame,
but something wild to run with.

— Robert Brault

In my discussion of Gary Snyder's ideas about wildness, I raised his
question about whether we could talk about cruelty in wild systems. I
also referred to the even more radical and disconcerting question raised
by the theologian Ernst Conradie as to whether animals can sin. These
are difficult questions, relating to the ethical (or otherwise) engagements
of animals with each other. As complex, maybe even more so, is the
exploration of human relationships with animals, which may in some
instances relate to the issue of how animals relate to each other. I cannot
pretend to offer clear answers to these questions, which have exercised
human societies for millennia. I simply want to offer some thoughts,
largely exploratory, which relate to the much humbler sense of what it
means to be human in a world that we share with innumerable other
species, on which we depend, and in ecological systems characterised
by multiple interdependencies, of which we frequently seem woefully
ignorant. I believe rethinking 'wildness' can be extremely useful in this
regard.

*   *   *

Homo sapiens have outgrown their use.
All the strangers came today,
And it looks as though they're here to stay.
　　　— David Bowie, 'Oh! You Pretty Things'

*　*　*

If the accounts of several visitors to the Kruger National Park are anything to go by, the perception that animals, especially predators, can be cruel is reasonably widespread. In the book *101 Kruger Tales: Extraordinary Stories from Ordinary Visitors to the Kruger National Park*, there are many such observations. Ursula Klitzke, for example, begins her account of wild dogs pursuing a bushbuck with the words, 'It's not always easy to understand the cruelty of nature', and then comments that they were grateful that the kill was concealed from their eyes by the thick bush 'as wild dogs are said to be the cruellest of hunters and will chase down their prey to exhaustion, often disembowelling the poor creature while it's still on its feet'.[1] Gabi Hotz, in turn, describes honey badgers as 'vicious little things'.[2] In her account of two visitors rescuing an impala, which was trapped in mud at a waterhole, from a long and painful death from hunger and thirst, or a certain end by predators, Wendy Abadi comments:

> I left the hide that afternoon knowing that those men were probably wrong for interfering. Nature is indeed cruel and the circle of life is, I suppose, as inevitable as the sun rising behind the Lebombos, or the sausage tree dropping its fruit. And yes, Kruger is not a zoo; tragic things will happen to helpless creatures and sometimes they will die in ways that seem neither fair nor humane. But, I must admit, I slept better that night knowing that the little *bokkie* was no longer stuck in the mud, waiting for thirst to consume it, or worse, for something to come along and rip it apart alive.[3]

In this chapter I begin by examining four poems that raise the ethics of animal-animal and human-animal relations issues in especially interesting ways. There are many poems that I could have chosen from different societies and different historical periods examining these relations, but these four seem to me to set out contrasting possibilities in intriguing ways, and have influenced my thinking on these matters over the years. The first is by the famous English poet, Ted Hughes, and is extremely well known, appearing on many high school literature syllabi. I quote it in its entirety for reasons that I hope will become apparent:

I sit in the top of the wood, my eyes closed.
Inaction, no falsifying dream
Between my hooked head and hooked feet:
Or in sleep rehearse perfect kills and eat.

The convenience of the high trees!
The air's buoyancy and the sun's ray
Are of advantage to me;
And the earth's face upward for my inspection.

My feet are locked upon the rough bark.
It took the whole of Creation
To produce my foot, my each feather:
Now I hold Creation in my foot

Or fly up, and revolve it all slowly –
I kill where I please because it is all mine.
There is no sophistry in my body:
My manners are tearing off heads –

The allotment of death.
For the one path of my flight is direct
Through the bones of the living.
No arguments assert my right:

The sun is behind me.
Nothing has changed since I began.
My eye has permitted no change.
I am going to keep things like this.
            (Ted Hughes, 'Hawk Roosting')

There is no doubt that Hughes captures the latent menace of the bird of prey at rest ('tearing off heads'), its perfect design as a top predator ('Between my hooked head and hooked feet'), and the fact that nothing escapes its penetrating eyesight from the height of its perch ('the earth's face upward for my inspection'). The poem is a remarkable act of the imagination, and captures, among many other things, the way in which, when faced with the penetrating gaze of a raptor from above, its speed, grace and self-sufficiency, one may frequently begin to wonder who holds authority in this encounter. With the 'sun behind me' – physically? as creator? as having itself been surpassed by the hawk? – the hawk is in complete control: 'Nothing has changed since I began. / My eye has permitted no change. / I am going to keep things like this.'

The response of the South African poet Douglas Livingstone was to dismiss Hughes's poem as 'anthropomorphic', in attributing human feeling and intention, albeit of a disturbing, menacing kind, to the hawk.[4] As both poet and scientist, Livingstone was insistent that animals do indeed have feelings ('If you're in any doubt, visit your local friendly abattoir,' he once commented), but felt that they needed to be registered differently. The charge of anthropomorphism is a complex one, perhaps more so than Livingstone's comment suggests, and I will return to this later.

But at this stage, I want to look at three of Livingstone's poems in which he himself reflects upon animals in situations of extreme suffering, and human responses to them, primarily his own: 'Gentling a Wildcat';[5] 'Bad Run at King's Rest';[6] and 'Carnivores at Station 22'.[7] 'Gentling a Wildcat' is the earliest of the poems, by date, and narrates an encounter between what is called in literary studies 'the speaker' – but let us assume it is Livingstone himself – and a wildcat in the process of giving birth, which has been mortally wounded in an encounter with a jackal:

> Under a tree, in filtered moonlight,
> a ragged heap of dusty leaves stopped moving.
> A cat lay there, open from chin to loins;
> lower viscera missing; truncated tubes
> and bitten-off things protruding.
> Little blood there was, but a mess of
> damaged lungs; straining to hold its breath
> for quiet; claws fixed curved and jutting,
> jammed open in a stench of jackal meat;
> it tried to raise its head hating the mystery, death.
> The big spade-skull with its lynx-fat cheeks
> aggressive still, raging eyes hooked in me, game;
> nostrils pulling at a tight mask of anger
> and fear . . .
> . . .
> Closely, in a bowl of unmoving roots,
> an untouched carcass, unlicked, swaddled and wrapped
> in trappings of birth, the first of a litter stretched.
> Rooted out in mid-confinement: a time
> when jackals have courage enough for a wildcat.

The desperate nature of the cat's injuries is presented vividly, down to the description of damaged organs, but Livingstone cannot provide the merciful release of death:

> In some things too, I am a coward,
> and could not here punch down with braced thumb,
> lift the nullifying stone or stiff-edged hand
> to axe with mercy the nape of her spine.

Something in him is deeply disturbed, not just by this death, but by the larger order of things of which it is a necessary part:

> And oppressively, something felt wrong:
> not her approaching melting with earth,
> but in lifetimes of claws, kaleidoscopes:
> moon-claws, sun-claws, teeth after death,
> certainly both at mating and birth.

His response is one of creaturely empathy, as he 'gentles' her:

> So I sat and gentled her with my hand,
> not moving much but saying things, using my voice;
> and she became gentle, affording herself
> the influent luxury of breathing –
> untrammelled, bubbly, safe in its noise.
> Later, calmed, despite her tides of pain,
> she let me ease her claws, the ends of the battle,
> pulling off the trapped and rancid flesh.
> Her miniature limbs of iron relaxed.
> She died with hardly a rattle.

And so the poet-scientist makes a gesture, perhaps of hopeless biological compassion, in placing her body and that of her cub out of the reach of predators, and willing into being the start of a gentler, more 'pastoral' cycle of life:

> I placed her peaceful ungrinning corpse
> and that of her firstborn in the topgallants
> of a young tree, out of ground reach, to grow: restart
> a cycle of maybe something more pastoral,
> commencing with beetles, then maggots, then ants.

The next two poems that I want to examine appeared in Livingstone's final collection before his untimely death, *A Littoral Zone*, a series of poems that emerged from his experience of testing the seawater at various points along the Durban coastline as part of his job as a marine bacteriologist. In the first poem, 'Bad Run at King's Rest', he encounters a stranded turtle, which has suffered fatal injury at human hands:

> The big loggerhead turtle lay
> swimming among human footprints, beached;
> shell split by an errant propeller-blade.
>
> Its flippers bloody where some lout's
> hacking had ripped nails for medicines
> or trophies. Both its eyes stabbed or pecked out.

Whether out of greater environmental disillusionment, or the extremity of the suffering of this creature, Livingstone does what he could not in 'Gentling a Wildcat':

> I moved – lifelong stand-in for thought –
> avoiding the still dangerous beak
> asking pardon, cut the leathery throat.

'Gentling a Wildcat' ends with some vain hope for a gentler natural order. In this poem, in which the inflicters of suffering are humans, there is no sense of redemptive possibility:

> Rinse off queasily. Circle wide,
> back, past that inert, spread-eagled mound.
> Call dumbly on gulls, on incoming tides.

'Carnivores at Station 22' poses even more searching questions about human-animal interactions. Here, Livingstone's calls for assistance from the local fishermen in his efforts to return a beached dolphin to the water are ignored. The fishermen are initially dehumanised, in that they appear at the outset 'poised like birds of prey / [who] stab sea and sky with bamboo beaks' (their rods). Their apparently silent indifference to the efforts to return the dolphin to the water reflects their contrary concern to that of Livingstone: that it is to them, not a sentient animal to whom they may have a moral responsibility, but instead a considerable source of food or bait:

> The men weighed these failed attempts
> – fishing for families, lean faces bland,
> the tide ebbing with my receding hopes –
> and looked away as I yelled and beckoned.

In this case, humans are not damned as individuals with moral consciences, but as a collective ('carnivores') and as a species ('I offered up a curse on *Homo sap*'), and the irony of the human response to the dolphin is pointed up in terms of a kind of biological morality, a moral code at the level of species, which the humans are about to violate, the dance image ('minuet') heightening the horror:

> The paradox has always haunted me:
> these are the only carnivores on earth
> that have never attacked man, yet
> out of their element they gain new worth.
>
> As I left, the fishermen stirred: stashing
> rods, moving in that ugly minuet
> of deliberate premeditation,
> one drawing a long and rusted bayonet.

These are troubling questions. Dolphins have very significant ecological roles to play in the maintenance of marine ecosystems, though they are inadvertently caught and often killed in the shark nets that protect bathers and surfers along the KwaZulu-Natal coast. The death of one dolphin will make little or no difference to the marine balance. And yet dolphins are beautiful, charismatic and intelligent creatures, and so most readers of this book would, like me, have leapt to Livingstone's assistance. The well-fed may afford compassion; but the faces of the fishermen are 'lean' and they are 'fishing for families'. Green and brown agendas clash in our responses to this scenario.

A similar dilemma was experienced recently, also on the KwaZulu-Natal coastline, when a large whale beached itself. It died, so there was no hope for it, but the images of local people rushing down with knives, pangas and machetes to harvest its meat caused much revulsion and disquiet in middle-class minds, perhaps my own too, I assume because many felt we should have been lamenting a tragedy rather than celebrating bounty. I recall a particular photograph on social media in which an African woman stands by the carcass, machete in hand, glaring back at the photographer, her face seeming to ask, 'Who are you to judge and photograph me when I am collecting food for my family?' The irony, of course, is that beached whale carcasses in KwaZulu-Natal are either loaded onto trucks to be dumped at a landfill site, or towed back out to sea to be consumed by sharks far from bathing areas, so the harvesting of the meat was to no one's detriment.

* * *

'See my works, how fine and excellent they are!' it says in Ecclesiastes, and the author, whether Kohelet the Preacher or King Solomon, was seldom wrong. 'All that I created, I created for you. Reflect on this and do not corrupt or desolate my world, for if you do, there will be no one to repair it after you.' — Esther Woolfson, Field Notes from a Hidden City

* * *

In a paper on the legacy of the apartheid state's failed project to cross-breed wolves with dogs, to produce a sort of 'super' attack dog, Louise Green argues that 'wild' is not a scientific category.[8] In one sense, she is absolutely correct, in that you cannot measure 'wildness' as you can genome sequences, for example. It is a quality, an understanding of a way of living, a set of expectations, and much more, including, as I pointed out in the case of the gorilla and the boy who fell into the zoo enclosure, assumptions about the ethics of interaction. But in another important sense, wildness *is* a scientific category, sometimes implicit and sometimes explicit: in many branches of science, indigeneity can only be attributed to something that is 'wild' – whose distribution is independent of human influence; and 'wildness' is also a measure used in conservation, in restricting or even prohibiting human access to areas of land or sea to preserve them as 'wilderness' or as 'reserves'. It is also, as the complaints of the fish farmers referred to earlier suggest, a problematic concept that is applied in the legislation of fish and animal species, even when they do not all exist in 'wild' populations.

In terms of human-animal interactions, there are a number of specific animals 'known' among animal trainers or those in the animal trade who traffic in unusual 'pets' to be so intrinsically 'wild' as to be untrainable, or unable to be tamed or domesticated (in 'Gentling a Wildcat', Livingstone comments, 'then I remembered hearing / they are quite impossible to tame'). The apartheid wolf-dog project, referred to above, which sought to create an attack dog that would have some of the characteristics of a wolf, 'its stamina, power, very resistant pads, superior coarse hair coat, stronger teeth, better heat resistance and its immunity to hip dysplasia',[9] apparently failed because the hybrids proved to be wild and 'untrainable'. Some species, such as birds like crows, starlings, mynahs or pigeons, are regarded as being readily domesticated. Others, such as the large cats, raptors, and several species of monkey and ape, are regarded as being able to be domesticated, but are never to be trusted, as they may reassert their 'wild' nature at any time. However complex and contradictory its meanings may be, 'wildness' remains a stubbornly insistent idea in understanding human-environmental and especially human-animal relations.

Whatever we may think of Hughes's poem 'Hawk Roosting', and Livingstone's criticism of it, it does remind us – in offering the hawk's perspective – that animals have their own consciousness. They look at us, they engage with us, as much as we look at them, engage with them. An exploration of human-animal interactions needs to understand this very significant fact. Wendy Woodward is a poet, literary critic and leading environmental studies scholar whose book, *The Animal Gaze: Animal Subjectivities in Southern African Narratives*, explores the complexities of human-animal interactions in fascinating new ways through this idea of animals looking, or looking back, at us.

Woodward argues that the way in which animals look at us 'compels a response' from us, because it contradicts any idea we may have of our superiority as humans over animals. We are not encountering an object, but rather an animal that claims and lives its own identity.[10] It is a reciprocal encounter, not simply one in which a human 'observes' something other. This is a position much more complex than the obvious observation that animals are creatures that (who? the pronoun is telling) live lives and respond to humans and other animals. Even those who routinely brutalise 'their' animals would likely admit as much (though probably with the qualifier that those 'lives' do not matter). Woodward allows that animals are 'beings' and 'identities' in their own right, with whom we have relationships that are complex and multidirectional. Those relationships also have important ethical dimensions, as do so many of our other relationships.

Woodward points out that Western philosophy has only recently been able to accept that an interchange of consciousnesses between humans and animals may be possible, and she quotes Peter Singer's observation that philosophers through much of the history of Western civilisation have thought of animals as 'beings of no ethical significance, or at best, of minor significance'.[11]

There are, of course, exceptions. The seventeenth-century philosopher John Locke stated: 'I think that people should be accustomed, from their cradles, to be tender to all sensible creatures, and to spoil or waste nothing at all.'[12] And attention to the rights of animals is evident in the thinking of eighteenth- and nineteenth-century philosophers such as Jeremy Bentham, John Stuart Mill and Henry Sidgwick, who 'insisted that the suffering of animals mattered in itself'.[13] I would, of course, add the notable contributions of the Romantic poet Percy Bysshe Shelley and the environmentalist, philosopher and poet Henry David Thoreau.

But Woodward points out, again via Singer, that even then 'human concerns have predominated over those of animals', and that the Western view of animals has only begun to change in the last 30 or so years:

[Philosophers from a variety of ethical traditions] have argued that the interests of animals deserve equal consideration with the similar interests of humans, or that animals have rights. They have sought to bridge the *ethical gap* that has hitherto been perceived to exist between members of our own species and members of other species.[14]

The major issue, Woodward argues, is whether humans can acknowledge 'subjective kinships' with animals, and what potential emerges if they can.[15] How would our views of, and behaviour towards, animals change? It is a move away from regarding humans as having a 'master consciousness'[16] that views and knows everything else, to one that acknowledges the interrelationships of knowing that characterise our interactions with other species.

What on earth does all of this have to do with 'wildness', some readers may ask? Behind just about all of the human activities that have damaged the planet's ecosystems, threatened species and, in many cases, obliterated them entirely, lies the assumption – formulated slightly differently at different times – that humans are the supreme species, that animals and plants are there for our own use and consumption, and that we have little ethical responsibility towards them. It is frequently presented in what I and many others see as a fundamental misunderstanding of the Christian creation narrative, in which all of non-human creation (except for that pesky apple tree . . .) is there for human consumption, enjoyment or use. An argument about 'wildness', or especially 'rewilding', is diametrically opposed to that understanding (but not to my own understanding of the biblical creation narrative). It removes humans from the centre of the argument, acknowledging our dependency on other species and theirs on us, and realising that relinquishing rather than asserting control over environments to allow biological processes to resume is frequently the most productive response.

In trying to move beyond what she sees as the limitations of much Western thinking about human-animal relations, Woodward turns to the beliefs and practices of a variety of indigenous societies, in which animals are understood to interact with humans on multiple levels, and share important aspects of identity. This happens especially, but not exclusively, in shamanic practices, such as, for example, Khoisan trance dances, in which healers believe they are transformed into animal-spirit identities. Much more fluid understandings of human-animal relations are widely evident in Khoisan belief systems, Aboriginal Dreaming, Native American ceremonies and storytelling, and the belief systems of numerous societies across the African continent, among many other examples. I have written about some of these myself, elsewhere.[17]

I do have two concerns about this line of argument, however. The first is that it can lead, in some cases, not necessarily Woodward's, to an idealised view that such societies operated 'in enlightened harmony with nature' and caused no environmental damage. George Monbiot makes it very clear that the environmental damage caused by such societies was often disproportionate to their size or apparent needs, and that more environmentally aware beliefs and practices were probably later developments. The second, related to the first, is the tendency for such belief systems about relations with the environment to be seen as unrelated to the specific histories in which they were practised, as if they represent an unchanging, timeless, 'natural' wisdom. Like those of all other human societies, the beliefs of hunter-gatherer, aboriginal or 'indigenous' societies (to pick some of the adjectives that appear in the literature) were subject to change and revision over time, often as a result of confrontation with environmental phenomena such as drought or fire, or brutal encounters with colonial occupation. It is, nevertheless, true that most are indeed characterised by much more complex, less hierarchical and more nuanced senses of human-animal interrelationships than much Western thinking offers in this regard, and can thus be instructive.

Woodward has her own concerns about some aspects of this thinking. She talks of the danger in ecopsychology and ecotourism of a tendency to try to expand one's own sense of identity by 'communing with' animals. She points out correctly, in my opinion, that ironically this becomes simply another way in which we assert the needs of humans over those of animals, in that the animals only have value to the extent that we can access them and identify with them.[18]

I said that I would return to the issue of anthropomorphism in human-animal relations, the charge laid by Livingstone against Hughes's representation of the hawk. Woodward refers to the work of several scholars who reject the charge of anthropomorphism on the basis that one *can* 'read' animals' emotional behaviour. Such scholars also point out that this charge is often the basis for rejecting the idea that animals are sentient beings. Perhaps more importantly for my purposes, she also refers to the argument of Marc Bekoff, an ethologist (someone involved in the scientific study of animal thinking), that anthropomorphism is in one sense 'an inevitable sin' because we only have human language to describe emotional lives and the experience of our own repertoire of emotions to refer to.[19] The perceptive reader will probably have noticed that even Livingstone, despite rejecting the anthropomorphic approach, describes the wildcat as 'hating the mystery, death', and as experiencing 'anger'. We may try to understand, but we can never actually know, for example, whether what appears to be fear in an animal is experienced as what we know as fear. Perhaps the most obvious possible misreading is that of dog owners who will judge that a reprimanded dog is displaying behaviour that attests to its 'guilt', whereas that behaviour is also that of the submissiveness of understanding his/her position in the hierarchy of the pack (which the various human and non-human members of the household effectively comprise). Bekoff does offer the necessary qualification that we must 'carefully, consciously, empathetically' try to 'maintain the animal's point of view', and seek to understand the experience of the individual living animal.[20]

Such experiences may apparently include pleasure. Woodward cites Jonathan Balcombe's argument that 'hedonic ethology' (studies that focus on pleasure in animal thinking) is extremely important in that it recognises animal consciousness and awareness, and 'contradicts the representation of nature as violent and unfeeling which "serves an insidious purpose: the continued exploitation of animals by humans".'[21] And in what may suggest that she has an answer to Snyder's or Conradie's questions at the beginning of this chapter about animal 'cruelty' or 'sin', Woodward again turns to Bekoff, who claims that '[t]he origins of virtue, egalitarianism and morality are more ancient than our own species'.[22]

One aspect of Woodward's argument that I do not follow, and to be fair it is more of an aside than a substantive line of discussion, is the claim that affinities between human and animal consciousnesses are supported by DNA. She claims that science itself is suggesting greater affinities between animals and humans, in that 'chimpanzees share 98.5% of their DNA with humans, and are thus closer to humans than they are to gorillas'.[23] I think one needs to be cautious about such claims, as we apparently also share 85 per cent of our DNA with mice, 61 per cent with fruit flies, and more than 50 per cent with bananas.[24]

In *Field Notes from a Hidden City: An Urban Nature Diary*, Esther Woolfson gives some attention to the question of moral behaviour in animals. I discussed what is called 'surplus killing' or 'henhouse syndrome' in the second chapter of this book as a troubling example of what might be gratuitous violence committed by predators. Woolfson denies outright that gratuitous violence is ever exhibited by animals: 'In wild creatures, violence is not purposeless. (We don't behave like "wild animals". If we behave badly, we behave as badly behaved humans.) There is always an explanation, even if we don't appreciate it or understand it. What we find difficult is to witness without judgement.'[25]

From all that I have read, the jury is still out on whether purposeless violence occurs in animals. Woolfson's argument does not help at all, though, because it is not actually an argument but a claim that she cannot back up. She simply says the phenomenon does not occur, and claims that there *is* a reason, even though we cannot find it, which flies in the face of logic.

Woolfson's discussion of moral behaviour in rats is more compelling. She quotes the work of Marc Bekoff and Jessica Pierce, in *Wild Justice: The Moral Lives of Animals*, who argue that laboratory tests suggest that 'rats will not take food if their doing so causes pain to other rats and that they will help other rats in distress – the motivation for doing so probably being empathy'.[26] She also cites the work of Claudia Rutte and Michael Taborsky, which indicates that rats 'who have been previously helped by other rats are more likely to offer help themselves', and another study that shows that rats 'being faced with two cages, one containing chocolate and one another rat, will quickly learn how to liberate their fellow before opening the second cage to share out the chocolate. The benefits bestowed upon the group by the cooperation of individuals is clear.'[27] Woolfson points out that studies at Cold Spring Harbor Laboratory in New York State have shown 'that rats presented with different types of visual and auditory stimuli are able to process and react to the information as efficiently as humans'.[28] This, however, leads her to ask the profoundly disturbing question as to why scientists at the same institution continue to use rats in their so-called 'forced swim test' or 'behavioural despair test', in which a rat is placed in a container of water from which it cannot escape, or the 'learned helplessness' test in which the animals are put through various forms of suffering so that scientists can measure at what stage they give up.[29] I find those tests almost impossible to comprehend, at every level. Woolfson asks, correctly, why the kind of stringent restrictions applied to the use of primates as research subjects are not extended to animals such as rats.

She also points out that birds that live close to humans can learn to distinguish between individual people, and that crows especially are able to identify individuals in a crowd who are a threat to them, apparently much to the dismay of researchers who have caught and ringed them or otherwise captured them for research purposes.[30]

As animals – birds in this case – learn to recognise individual human beings, so can we frequently identify individuals in other species, perhaps learn some of their distinctive ways of behaving, and frequently give them names, as Woolfson does with the numerous creatures with whom she shares her house. Naming individual animals is a contradictory impulse. It may imply individual recognition, value, familiarity, an awareness of distinctiveness and 'character'. It also frequently suggests 'ownership', an assertion of power, and sometimes the authority to discipline. The contradictions are especially apparent in the practice of naming 'wild' animals in game parks, which may endear them to visitors (an important conservation function), but also clearly identifies them as subject to the ownership and control of the park owners rather than 'wild' inhabitants of an ecosystem.

* * *

Take me back to the days of the foreign telegrams
and the all-night rock and rollin' . . . hey Shell
we was wild then. — Michelle Shocked, 'Anchorage'

* * *

Does an argument about taking seriously the ethics of our interactions with animals lead one inevitably down the road to vegetarianism? The rationales for vegetarianism are so historically and individually varied that one cannot even attempt an answer. Among so many others, they range from: aversion to killing animals; the perceived digestive 'heaviness' of meat; distaste for animal products as 'unclean' (think of the Romantic poet Shelley's comment: 'It is only by softening and disguising dead flesh by culinary preparation that it is rendered susceptible of mastication or digestion, and that the sight of its bloody juices and raw horror does not excite intolerable loathing and disgust');[31] vegetarianism as conducive to meditation; the expense of meat; vegetarianism or veganism as perceived to be more environmentally sustainable lifestyles; animals being regarded as sacred.

What is clear, though, is that an alertness to animals as sentient creatures, which as Matthew Scully suggests 'cannot ask us for mercy',[32] places a significant burden on us as humans to ensure their best possible care, and – if we choose to eat them or utilise their by-products – the most humane possible mode of killing them. (It is perhaps telling that the Constitution of South Africa, which is widely regarded as one of the most progressive and exemplary in the world, offers no protection to animals.) Of whatever else it may be guilty, the hunting lobby is frequently unfairly labelled as 'barbaric' by environmental activists. In my view, a clean shot to the heart or brain of an antelope is probably the most humane way in which to kill it, certainly infinitely preferable to a cramped ride on the back of a truck to queue for the abattoir. It is a philosophy espoused, for example, by the chef Gordon Wright, who serves in his restaurant in the town of Graaff-Reinet venison and game birds that he has himself hunted. He regards this as an ethical and health imperative, in that he is able 'to shoot a buck so cleanly it feels nothing. And there have been no feedlots, no steroids, no stress.'[33] I think such an awareness also involves, as far as possible, an attitude of gratitude, and a commitment not to waste, as argued by well-known angling author Harold F. Blaisdell: 'One must not waste, or be careless, with the bodies or the parts of any creature one has hunted. One must not boast, or show much pride in accomplishment, and one must not take one's skill for granted.'[34] After all, an animal has given up its life for us.

The reader may, at this stage, legitimately turn my own line of argument back on me, as I have provided detailed accounts of my fishing, and ask about my ethical view on that. I have to admit, honestly, to feeling conflicted. For years I simply followed the line of argument that fish had limited capacity for feeling pain or distress, which suggested that fishing was not especially 'cruel'. Evidence for their lack of capacity for pain was adduced by the fact that many of them eat prey items (bees, whole [live] crabs, mud prawns, spiny finned baitfish, mussels that are crushed whole, and so on), which would be off limits were their mouths, and stomachs, anything like as receptive to pain as ours. The fact that many of us now release the bulk of our catch also seemed to testify to the ethical nature of the engagement. Many of those who choose vegetarianism for ethical reasons seemed to follow this line of argument, in deciding that meat and poultry were unacceptable to eat, but fish was not as fish did not suffer in being caught and killed in the same way as birds and animals (Woodward follows this line of argument in calling herself a pescatarian).

But new research suggests a more complex picture, which casts much of this into question. In particular, Jonathan Balcombe's book, *What a Fish Knows: The Inner Lives of our Underwater Cousins*,[35] presents a much more complex view of the thought processes, physical experiences, emotional lives and characters of fish than was ever imagined. If Balcombe is correct that fish experience distress and pain in being caught, then catch and release, while undoubtedly a good conservation strategy, may be the most ethically problematic way to proceed: fish being made to suffer for no purpose other than the personal enjoyment of the angler. Perhaps killing a fish or two for the table and then folding up the rod is the more 'honest' way to proceed. That approach may, however, work on stillwaters with fairly large populations of fish, but would quickly destroy the populations of fish in the smaller rivers. In all of this, I also know that the closest friends and defenders of river environments are those who fish them most often and extensively, and so are the most alert to any possible degradation of the ecosystem. As I say, I remain conflicted on this issue, but have also come to another realisation: that I am increasingly uncomfortable with fisheries that are under such severe fishing pressure that the fish are daily harried by anglers, and so are unable to live lives on anything like their 'own' terms; rather, they appear to exist only for the sake of humans, who pursue them relentlessly. Rewilding one's life is a process, and mine may yet take me in directions I had never anticipated.

10

A Wilder Mind

I had been blessed with a wilder mind.
— Mumford and Sons, 'A Wilder Mind'

In one sense, this book can have no conclusion for it cannot tie up its arguments neatly, nor can it declare an end to what must for each of us be the work of a lifetime, and cumulatively for our species the work of multiple lifetimes. In another sense, it must have a conclusion, a checking of where we have come to on this journey of arguments, narratives and reflections, a tightening of the straps on the load, before we continue.

I completed this book while Cape Town was experiencing a drought so severe that water supplies were intermittently turned off, and the distribution of drinking water by armed security forces was being contemplated. At the same time, Durban had just experienced such a dramatic storm and flooding in one afternoon that fifteen people had died, roofs and cars had been tossed around like twigs, and several areas had been declared disaster areas. Even the climate change sceptics seem, at least temporarily, to have been silenced.

I have the strongest aversion to the tones of castigation, blame and threat that characterise so many discussions of climate change – no one really appreciates being lectured at – and which I suspect do little other than antagonise the people they are claiming to reach while they salve the consciences of those at the lectern, keyboard or microphone.

I hope that, in contrast, the journey that the reader has shared with me in this book is one of possibility, affirmation and excitement; that it suggests ways of thinking, acting, speaking, which evince the greatest environmental responsibility and commitment to the sustainability of global environments, while they simultaneously expand the possibilities and pleasures of how we live. It is, finally, a hopeful journey, despite the ugly facts that we have to admit about our past and present conduct as humans.

Although he does not feature as prominently as many other authors in this book, perhaps my most significant encounter has been with Henry David Thoreau, a thinker who has radically expanded my sense of what a life may be, while at the same time rapidly shrinking my understanding of what such a life requires. Reflecting on the fact that many people lead lives of 'quiet desperation'[1] in their pursuit of what they or their society assume is required for a fulfilled or 'happy' life, Thoreau comments:

> Most of the luxuries, and many of the so called comforts of life, are not only not indispensable, but positive hindrances to the elevation of mankind. With respect to luxuries and comforts, the wisest have ever lived a more simple and meager life than the poor. The ancient philosophers, Chinese, Hindoo, Persian, and Greek, were a class than which none has been poorer in outward riches, none so rich inward.[2]

(As an aside, is the band Pink Floyd quoting Thoreau in the line from *Dark Side of the Moon*, 'Hanging on in quiet desperation is the English way'?) I am far too fond of a good glass of wine, a decent whisky, or a medium-rare steak to opt for the rice, water and unleavened bread regimen that Thoreau espouses in his experiment in living the simplest life possible, recounted in his book *Walden*. I am, however, intrigued by the concept of a life stripped of what are regarded as pleasures, but the pursuit of which actually becomes anything but pleasurable in the endless round of effort we put into securing them. I could never follow

Thoreau's extreme asceticism, which makes Robert Baden-Powell look positively decadent, but I am inspired to pause at the thought of six weeks of work per year being sufficient to cover one's basic needs, and to allow dramatically expanded opportunities for leisure and/or study.

Thoreau's model of living very lightly on the earth is one that is echoed in many ways by thinkers or ascetics from various societies at different points in history, and it is suggestive in a global context in which natural resources are hopelessly insufficient for the model and dream of American middle-class consumption to be extended to all inhabitants. Expectations about what is an appropriate level of consumption simply need to be adjusted down radically, for everyone, if we are to survive as a species or a set of interdependent species in one giant ecosystem called Earth. One could argue that Africa's underdevelopment may ironically turn out to be its saving grace. But that is not an easy argument to make, for example, in a socio-economic context like South Africa, in which radical inequality means that over 50 per cent of the population is living in conditions of enforced deprivation. I would not relish making the argument to such people that any aspiration they may have for the middle-class consumer lifestyle celebrated in television series and magazines is entirely inappropriate because the environment cannot sustain it, even though it is undoubtedly true. People who have queued for centuries finally to get to the door will not be impressed to be told by the doorman that the banquet cannot feed any more people. Green and brown agendas continue to jostle each other.

I do not find grand, overarching proclamations terribly helpful in addressing the kind of issues raised in this book, though there is clearly need for concerted global action on matters such as industry standards for emissions, the protection of marine environments and many other problems that are beyond the remit of one country, which makes the utter ignorance and arrogance of Donald Trump's environmental denialism possibly the biggest threat we have to face currently. What interests me more are modes of localised action and monitoring, through which local inhabitants and their environments experience an improvement in the quality of their lives, and an enlarged sense of what it actually means to live. I was struck when, at a recent conference, the well-known Humanities scholar, Homi K. Bhabha, said in discussion after a paper he had given: 'When you live in a region of seismic instability, you build in bamboo.' It occurred to me that in the complexities of negotiating these issues, we do not need monolithic arguments like great steel structures imposed from above, but flexible, movable arguments adapted to their locations and purposes.

In the process of conducting the research for this book, I read Jon Krakauer's *Into the Wild*, in which he narrates the story of the life and death of Chris McCandless, and his quest for another kind of life in walking alone, after many other similar ventures, into the wilderness of Alaska. I have no desire whatsoever to follow his model, but I am still haunted by this book on an almost daily basis, in McCandless's desire and will to explore the possibilities of life lived to its fullest and sparsest. I think those who focus on the foolishness of what he did, or claim to be able to point to 'obvious' mistakes that he made through apparent lack of local knowledge, are entirely missing the point: that of a complex and intelligent person driven to seek a life pared down to its thinnest and purest, but also its richest, despite the extreme discomfort and danger involved; driven to find something infinitely greater in far, far less.

What does not interest me in this regard, though I do watch them for entertainment value, are the television reality series featuring various people who have moved to remote locations across the globe to become modern hunter-gatherers. Despite the availability of good camping gear and cheap ammunition in the US, which is where most of them seem to originate, they appear to create lives of extremely uncomfortable faux 'primitivism' in wearing animal skins, sleeping in poorly waterproofed shelters, and affecting body markings that are a strange jumble of 'tribal' symbolism from very different societies, even though it frequently becomes evident in the programme that there is a store nearby, which supplies the usual necessities, such as clothes, gas and tools. Some of them secure their food or livelihood by trapping animals, which is an inhumane practice I detest.

As I have argued in this book, I do not see 'rewilding' as involving an abandonment of urban life, agriculture, technology, health care, and so on, though it may frequently require that we view them differently. But I do understand something of the impulse behind the lifestyle choices of those featured on such programmes: the desire for a more 'connected' life. And I insist that the only way to get people to embark upon courses of action to halt climate change and environmental damage is to let them see and feel the benefits to themselves. When I first read George Monbiot's *Feral*, I was rather perplexed by his diagnosis of himself as suffering from 'ecological boredom' to which 'rewilding' was a powerful antidote. I could almost understand it intellectually, but it did not resonate particularly strongly for me on a personal level. Now, as I reach the end of my own account, I realise how much of my life, from my earliest years, has involved 'straying back into the woods', to borrow Gary Snyder's suggestive phrase from the epigraph to this book, as a response to a restiveness perhaps induced indeed by 'ecological boredom'. I am struck, too, by how significantly – in the process of researching and writing this book – my life has been invigorated, expanded, reoriented by thinking with, through, about, into 'the wild' and 'wildness', and the quite extraordinary authors, people, places, creatures and encounters that have both mapped and marked the journey. And I feel I have been 'blessed with a wilder mind'.

Notes

Chapter 1: Into the Wild

1. Holmes Rolston III, *Environmental Ethics: Duties to and Values in the Natural World* (Philadelphia: Temple University Press, 1988), p. 3.
2. Rolston, *Environmental Ethics*, p. xi.

Chapter 2: Rewilding

1. Scott Ramsay, '9 of the Wildest Experiences in Southern Africa?', *Getaway* 26(10) (January 2015), p. 46.
2. Quoted in Peter B. Landres, Mark W. Brunson, Linda Merigliano, Charisse Sydoriak and Steve Morton, 'Naturalness and Wildness: The Dilemma and Irony of Managing Wilderness', *USDA Forest Service Proceedings* (RMPRS-P-Vol-5) (2000), p. 377.
3. Landres et al., 'Naturalness and Wildness', p. 377.
4. See Duncan Brown, *Are Trout South African? Stories of Fish, People and Places* (Johannesburg: Picador Africa, 2013), p. 22.
5. Landres et al., 'Naturalness and Wildness', p. 377.
6. Landres et al., 'Naturalness and Wildness', p. 378, quoting Andres et al.
7. Landres et al., 'Naturalness and Wildness', p. 378.
8. Landres et al., 'Naturalness and Wildness', p. 378.
9. Landres et al., 'Naturalness and Wildness', p. 379.
10. Landres et al., 'Naturalness and Wildness', p. 380.
11. Landres et al., 'Naturalness and Wildness', p. 381.
12. George Monbiot, *Feral: Rewilding the Land, Sea and Human Life* (London: Penguin, [2013] 2014).
13. Ben Ridder, 'The Naturalness versus Wildness Debate: Ambiguity, Inconsistency, and Unattainable Objectivity', *Restoration Ecology* 15(1) (2007): 8.
14. Ridder, 'The Naturalness versus Wildness Debate': 9.
15. Ridder, 'The Naturalness versus Wildness Debate': 9.

16. Monbiot, *Feral*, p. 69.

17. See Brown, *Are Trout South African?* in this regard.

18. Ridder, 'The Naturalness versus Wildness Debate': 9, quoting Oliver at al.

19. Ridder, 'The Naturalness versus Wildness Debate': 9.

20. Ridder, 'The Naturalness versus Wildness Debate': 9, quoting Flannery.

21. Ridder, 'The Naturalness versus Wildness Debate': 10.

22. Ridder, 'The Naturalness versus Wildness Debate': 10.

23. Harold F. Blaisdell, *The Philosophical Fisherman: Reflections on Why We Fish* (New York: Skyhorse Publishing, [1969] 2015), p. 167.

24. Gary Snyder, *The Practice of the Wild* (Berkeley: Counterpoint, 1990), p. 6.

25. Snyder, *The Practice of the Wild*, p. 2.

26. Snyder, *The Practice of the Wild*, p. 9.

27. Snyder, *The Practice of the Wild*, pp. 9–11.

28. Snyder, *The Practice of the Wild*, p. 193.

29. Snyder, *The Practice of the Wild*, pp. 193–4.

30. Snyder, *The Practice of the Wild*, p. 192.

31. Gary Snyder and Jim Harrison, *The Etiquette of Freedom and the Practice of the Wild*, edited by Paul Ebenkamp (Berkeley: Counterpoint, 2010), p. 12.

32. Brown, *Are Trout South African?*, p. 22.

33. Snyder, *The Practice of the Wild*, p. 16.

34. Snyder, *The Practice of the Wild*, p. 76. It is an observation akin to that of theologians who have remarked on the irony that the only biblical injunction humans seem faithfully to have obeyed is to 'Go forth and multiply' (Genesis 1: 28).

35. Snyder, *The Practice of the Wild*, p. ix.

36. Snyder, *The Practice of the Wild*, p. 17.

37. Snyder, *The Practice of the Wild*, pp. viii–ix.

38. Snyder, *The Practice of the Wild*, p. 9.

39. Snyder, *The Practice of the Wild*, p. 7.

40. Snyder, *The Practice of the Wild*, pp. 15–16.

41. Snyder, *The Practice of the Wild*, p. 12.

42. Snyder and Harrison, *The Etiquette of Freedom*, p. 11.

43. Snyder, *The Practice of the Wild*, pp. 99–100.

44. Snyder, *The Practice of the Wild*, p. viii.

45. Snyder, *The Practice of the Wild*, p. x.

46. Snyder, *The Practice of the Wild*, p. 22.

47. Snyder, *The Practice of the Wild*, p. 24.

48. Snyder, *The Practice of the Wild*, p. 25.

49. Snyder, *The Practice of the Wild*, p. 127.

50. Snyder, *The Practice of the Wild*, p. 101.

51. Snyder, *The Practice of the Wild*, p. 26.

52. Snyder, *The Practice of the Wild*, p. 101.

53. Snyder, *The Practice of the Wild*, p. 99.

54. Snyder, *The Practice of the Wild*, p. 97.

55. Ernst Conradie, 'Do Humans Sin? In Conversation with Frans de Waal', Paper delivered at a conference of the European Society for the Study of Science and Theology on the theme 'Are We Special? Science and Theology on Human Uniqueness' (Warsaw, 2016).

56. Snyder, *The Practice of the Wild*, p. 118.

57. Monbiot, *Feral*, p. 114.

58. 'Surplus Killing', 2016, http://en.wikipedia.org/wiki/Surplus_killing (accessed 20 May 2016).

59. I recall attending a colloquium on environmental issues at which the main meal promised to be 'Waterblommetjie Bredie', usually a delicious lamb stew cooked with the buds of a plant similar to a water lily. This one seemed alarmingly literal in its interpretation of the name of the dish: it comprised boiled waterblommetjies.

60. Snyder, *The Practice of the Wild*, p. 197.

61. Snyder, *The Practice of the Wild*, p. 197.

62. Snyder, *The Practice of the Wild*, p. 20.

63. Snyder, *The Practice of the Wild*, pp. 43, 45, 127.

64. Snyder, *The Practice of the Wild*, p. 45 and *passim*.

65. Monbiot, *Feral*, p. 86.

66. Roderick Haig-Brown, *Fisherman's Summer*, second edition (Ontario: Totem Books, [1959] 1975), pp. 165–73.

67. Monbiot (*Feral*, p. 8) notes that the term 'rewilding' entered the Chambers dictionary in 2011.

68. Monbiot, *Feral*, pp. 7–8.

69. Monbiot, *Feral*, p. 198.

70. Monbiot, *Feral*, p. 138.

71. Monbiot, *Feral*, p. 8.

72. Monbiot, *Feral*, p. 8.

73. Monbiot, *Feral*, p. 11.

74. Monbiot, *Feral*, p. 8.

75. Monbiot, *Feral*, p. 8.

76. Monbiot, *Feral*, p. 9.

77. Paul Schullery, *Royal Coachman: Adventures in the Fly Fisher's World* (Albuquerque: University of New Mexico Press, [2000] 2007).

78. Monbiot, *Feral*, p. 9.
79. Monbiot, *Feral*, p. 91.
80. Monbiot, *Feral*, pp. 9–10.
81. Monbiot, *Feral*, p. 105.
82. Monbiot, *Feral*, p. 10.
83. Monbiot, *Feral*, p. 87.
84. Monbiot, *Feral*, p. 10.
85. Monbiot, *Feral*, pp. 10–11.
86 Monbiot, *Feral*, p. 11.
87. Monbiot, *Feral*, p. 12.
88. Monbiot, *Feral*, p. 196.
89. Monbiot, *Feral*, p. 208.
90. Monbiot, *Feral*, p. 120.
91. Monbiot, *Feral*, p. 115.
92. Monbiot, *Feral*, pp. 124–3.
93. Monbiot, *Feral*, p. 67.
94. Monbiot, *Feral*, p. 69.
95. 'Rare Plant "Treasure" at Vergelegen Estate', *Bolander* (Wednesday 10 August), p. 10.
96. Monbiot, *Feral*, p. 155.
97. Monbiot, *Feral*, p. 70.
98. Monbiot, *Feral*, p. 158.
99. Monbiot, *Feral*, pp. 154–5.
100. Monbiot, *Feral*, p. 84.
101. Monbiot, *Feral*, p. 84.
102. Monbiot, *Feral*, p. 84.
103. Monbiot, *Feral*, p. 85. Yellowstone is also discussed extensively in this respect by Schullery, *Royal Coachman*.
104. Monbiot, *Feral*, p. 142.
105. Monbiot, *Feral*, p. 144.
106. Monbiot, *Feral*, p. 145.
107. Monbiot, *Feral*, p. 256.
108. Monbiot, *Feral*, p. 32–3.
109. Monbiot, *Feral*, p. 33.
110. Monbiot, *Feral*, p. 33.
111. Monbiot, *Feral*, p. 3.

Chapter 3: Wildness and Conservation

1. All sold under the brand name 'Safari', though by different companies.
2. Andrea Abbott, 'Return of the Wild', *South African Country Life* (June 2017), p. 41.
3. Abbott, 'Return of the Wild', p. 44.
4. Marja Spierenburg and Shirley Brooks, 'Private Game Farming and its Social Consequences in Post-apartheid South Africa: Contestations over Wildlife, Property and Agrarian Futures', *Journal of Contemporary African Studies* 32(2) (2014): 151.
5. Spierenburg and Brooks, 'Private Game Farming and its Consequences': 156.
6. Spierenburg and Brooks, 'Private Game Farming and its Consequences': 152.
7. Spierenburg and Brooks, 'Private Game Farming and its Consequences': 153.
8. Spierenburg and Brooks, 'Private Game Farming and its Consequences': 154.
9. Spierenburg and Brooks, 'Private Game Farming and its Consequences': 155.
10. Spierenburg and Brooks, 'Private Game Farming and its Consequences': 160.
11. Spierenburg and Brooks, 'Private Game Farming and its Consequences': 165.
12. Femke Brandt and Marja Spierenburg, 'Game Fences in the Karoo: Reconfiguring Spatial and Social Relations', *Journal of Contemporary African Studies* 32(2) (2014): 221.
13. Brandt and Spierenburg, 'Game Fences in the Karoo': 229.
14. Brandt and Spierenburg, 'Game Fences in the Karoo': 220. In a comment on an earlier draft of this chapter, Antjie Krog remarked that she was surprised that the shift to game farming was not discussed more frequently as a response to crime, with stock theft driving many farmers to have abandoned sheep farming some years ago, and more recently even cattle.
15. Brandt and Spierenburg, 'Game Fences in the Karoo': 232.
16. At the time of writing, there was significant public debate about the ANC's plans to amend the South African Constitution to allow for the expropriation of land without compensation, but the amendment was still at the level of proposal.
17. Brandt and Spierenburg, 'Game Fences in the Karoo': 234.
18. Brandt and Spierenburg, 'Game Fences in the Karoo': 226.
19. Brandt and Spierenburg, 'Game Fences in the Karoo': 226.
20. Njabulo Ndebele, *Fine Lines from the Box: Further Thoughts about Our Country* (Cape Town: Umuzi, 2007), pp. 101–2.
21. Ndebele, *Fine Lines from the Box*, p. 105.
22. Duncan Brown, *Are Trout South African? Stories of Fish, People and Places* (Johannesburg: Picador Africa, 2013), pp. 39–40.
23. I am indebted to Peter Brigg's account of this project in 'The River King of Thendela', *South African Country Life* (July 2015), pp. 82–5.

24. Knut G. Nustad, *Creating Africas: Struggles over Nature, Conservation and Land* (London: Hurst, 2015), p. 7.
25. Nustad, *Creating Africas*, p. 7.
26. Nustad, *Creating Africas*, pp. 7–8.
27. Nustad, *Creating Africas*, p. 8.
28. Nustad, *Creating Africas*, p. 17.
29. Nustad, *Creating Africas*, pp. 9, 14.
30. Nustad, *Creating Africas*, p. 13.
31. Nustad, *Creating Africas*, p. 19.
32. I have argued about the importance of trout in this regard for colonial settlers in South Africa, and about the fact that one should not try to draw too clear a distinction between fishing for food or sport in such contexts (Brown, *Are Trout South African?*).
33. Nustad, *Creating Africas*, pp. 22–3.
34. Nustad, *Creating Africas*, pp. 23–4.
35. Nustad, *Creating Africas*, p. 24.
36. Nustad, *Creating Africas*, p. 26.
37. Jane Carruthers, *The Kruger National Park: A Social and Political History* (Pietermaritzburg: University of Natal Press, 1995), p. 100.
38. Elsie Cloete, '"There's a Meat down There": An Essay on English and the Environment in Africa', in *Vernacular Worlds, Cosmopolitan Imagination*, edited by Stephanos Stephanides and Stavros Karayanni (Leiden and Boston: Brill and Rodopi, 2015), p. 33.
39. Nustad, *Creating Africas*, pp. 27–8.
40. Nustad, *Creating Africas*, p. 28.
41. Nustad, *Creating Africas*, p. 28.
42. Nustad, *Creating Africas*, p. 30.
43. Quoted by Nustad, *Creating Africas*, p. 30.
44. Nustad, *Creating Africas*, p. 31.
45. Nustad, *Creating Africas*, p. 38.
46. Quoted by Nustad, *Creating Africas*, p. 58.
47. Ingold, quoted by Nustad, *Creating Africas*, p. 58.
48. Jeff Gordon (ed.), *101 Kruger Tales: Extraordinary Stories from Ordinary Visitors to the Kruger National Park* (Cape Town: Leadwood Publishing, 2014), p. 247.
49. Gary Snyder, *The Practice of the Wild* (Berkeley: Counterpoint, 1990), p. 30.
50. Quoted in Gregory H. Aplet, 'On the Nature of Wildness: Exploring What Wilderness Really Protects', 1998–1999, http://heinonline.org/HOL/LandingPage?handle=hein.journals/denlr76&div=19&id=page= (accessed 12 September 2011), p. 361.
51. Quoted in Aplet, 'On the Nature of Wildness', p. 350.

Chapter 4: Wildness and Local Language

1. Duncan Brown, *To Speak of this Land: Identity and Belonging in South Africa and Beyond* (Pietermaritzburg: University of KwaZulu-Natal Press, 2006), pp. 7–35.
2. Chris Robinson and Eileen Finlayson, *Scottish Weather* (Edinburgh: Black and White, 2008). There are any number of websites listing these terms and their meanings.
3. George Monbiot, *Feral: Rewilding the Land, Sea and Human Life* (London: Penguin, [2013] 2014), pp. 49, 65, 150.
4. Robert Macfarlane, 'The Word-Hoard: Robert Macfarlane on Rewilding Our Language of Landscape', 27 February 2015, www.theguardian.com/books/2015/feb/27/robert-macfarlane-word-hoard-rewilding-landscape (accessed 27 February 2015), p. 1.
5. Macfarlane, 'The Word-Hoard', p. 1.
6. Macfarlane, 'The Word-Hoard', p. 3.
7. Macfarlane, 'The Word-Hoard', p. 1.
8. Macfarlane, 'The Word-Hoard', p. 3.
9. Macfarlane, 'The Word-Hoard', p. 3.
10. Macfarlane, 'The Word-Hoard', p. 1.
11. Macfarlane, 'The Word-Hoard', p. 2.
12. Macfarlane, 'The Word-Hoard', p. 2.
13. Macfarlane, 'The Word-Hoard', p. 2.
14. Macfarlane, 'The Word-Hoard', p. 1.
15. Macfarlane, 'The Word-Hoard', p. 2.
16. Macfarlane, 'The Word-Hoard', p. 4.
17. Howard Poynton, 'The Smell of Rain: How CSIRO Invented a New Word', *The Conversation* (31 March 2015), http://theconversation.com/the-smell-of-the-rain-how-csiro-invented-a-new-word-39231 (accessed 23 January 2017).
18. Poynton, 'The Smell of Rain', p. 1.
19. Poynton, 'The Smell of Rain', p. 2.
20. Poynton, 'The Smell of Rain', p. 2.
21. Leon de Kock, 'The Land and its Appropriation by "English"', in *Literature, Nature and the Land: Ethics and Aesthetics of the Environment* (Collected AUETSA Papers 1992), edited by Nigel Bell and Meg Cowper-Lewis (Ngoye: University of Zululand, 1993), p. 208.
22. Jean Comaroff, *Body of Power, Spirit of Resistance: The Culture and History of a South African People* (Chicago: University of Chicago Press, 1985), p. 138.
23. R.H.W. Shepherd, *Lovedale South Africa: The Story of a Century, 1841–1941* (Lovedale: Lovedale Press, 1940), p. 67.

24. Quoted in De Kock, 'The Land and its Appropriation by "English"', p. 209.
25. I am grateful to Jeff Opland for this information (personal communication).
26. I am grateful to Antjie Krog and Stephen Boshoff for checking my translations and making suggestions.
27. Elsie Cloete, '"There's a Meat down There": An Essay on English and the Environment in Africa', in *Vernacular Worlds, Cosmopolitan Imagination*, edited by Stephanos Stephanides and Stavros Karayanni (Leiden and Boston: Brill and Rodopi, 2015), p. 26.
28. Cloete, '"There's a Meat down There"', p. 27.
29. Cloete, '"There's a Meat down There"', p. 27.
30. Cloete, '"There's a Meat down There"', p. 27.
31. Cloete, '"There's a Meat down There"', p. 28.
32. Cloete, '"There's a Meat down There"', p. 28.
33. Cloete, '"There's a Meat down There"', p. 29.
34. Cloete, '"There's a Meat down There"', p. 29.
35. Lee Gutteridge and Louis Liebenberg, *Mammals of Southern Africa and their Tracks and Signs* (Johannesburg: Jacana, 2013), p. 28.
36. Gutteridge and Liebenberg, *Mammals of Southern Africa*, pp. 29–47.
37. Cloete, '"There's a Meat down There"', p. 30. She draws this from William Wolmer, *From Wilderness Vision to Farm Invasions: Conservation and Development in Zimbabwe's Southeast Lowveld* (Oxford: James Currey, 2007), p. 45.
38. Cloete, '"There's a Meat down There"', p. 30.
39. Cloete, '"There's a Meat down There"', p. 31.
40. Cloete, '"There's a Meat down There"', p. 34.
41. Cloete, '"There's a Meat down There"', p. 35.
42. W.H.I. Bleek and L.C. Lloyd, *Specimens of Bushman Folklore* (London: George Allen and Company, 1911), pp. 251–3.

Chapter 5: Wild Seams and Margins

1. Brett Archibald, *Alone: The Search for Brett Archibald* (Johannesburg: Jacana, 2016), p. 64.
2. Archibald, *Alone*, p. 18.
3. Archibald, *Alone*, p. 31.
4. My translation.
5. Douglas Livingstone, *The Anvil's Undertone* (Johannesburg: Ad Donker, 1978), pp. 39–40.
6. Henrietta Rose-Innes, *Green Lion* (Cape Town: Umuzi, 2015), p. 21.
7. Rose-Innes, *Green Lion*, p. 23.

8. 'Crime-Fighting Gorilla Dies', *News24*, 5 May 2004, http://news24.com/SouthAfrica/News/Crime-fighting-gorilla-dies-20040505 (accessed 16 January 2017).

9. Mojalefa Mashego and Gill Gifford, 'Max the Gorilla's Shooter's Riddle'. *Independent Online*, 4 November 2005, http://www.iol.co.za/news/south-africa/max-the-gorilla-shooters-death-riddle-257947 (accessed 16 January 2017), p. 1.

10. Though the physician Ray Melamed, who was a fellow at the Stellenbosch Institute for Advanced Study while I was there working on this project, told me of an experiment in which mice were trained to regulate their own blood pressure/heart rate through receiving micro-shocks when these became too high.

11. Douglas Livingstone, *A Littoral Zone* (Cape Town: Carrefour Press, 1991), p. 7.

Chapter 6: Wild Cities

1. See https://www.theguardian.com/cities/2018/jul/23/darwin-comes-to-town-how-cities-are-creating-new-species (accessed 2 August 2018).

2. Sarah Emerson, 'Boston is Covered in Goose Poop and People are as Mad as Hell', 25 October 2006, http://motherboard.vice.com/read/boston-is-covered-in-goose-poop-and-people-are-mad-as-hell (accessed 28 October 2016).

3. I am grateful to Anel Pieterse for alerting me to this.

4. Esther Woolfson, *Field Notes from a Hidden City: An Urban Nature Diary* (London: Granta, 2014), pp. 5–6.

5. Woolfson, *Field Notes from a Hidden City*, p. 6.

6. Woolfson, *Field Notes from a Hidden City*, pp. 6–7.

7. Woolfson, *Field Notes from a Hidden City*, p. 8.

8. Woolfson, *Field Notes from a Hidden City*, p. 69.

9. Woolfson, *Field Notes from a Hidden City*, p. 65.

10. George Monbiot, *Feral: Rewilding the Land, Sea and Human Life* (London: Penguin, [2013] 2014), pp. 49–61.

11. Monbiot, *Feral*, pp. 50–1.

12. Abby-Gene Bissolati, 'Luiperd "Hou Vakansie" in Huis in Gordonsbaai', *Die Burger*, 10 October 2016, pp. 1, 9.

13. The information that follows is drawn from the documentary.

14. I am indebted to Don Pinnock, *Wild as It Gets: Wanderings of a Bemused Naturalist* (Cape Town: Tafelberg, 2016), pp. 27–30, for his account of this research.

15. Pinnock, *Wild as It Gets*, p. 29.

16. Ted Hughes, *Selected Poems 1957–1981* (London: Faber and Faber, 1984), p. 41.

17. Henry David Thoreau, *Walden* (London: Penguin, [1854] 2016), p. 217.

18. Woolfson, *Field Notes from a Hidden City*, p. 207.

19. Woolfson, *Field Notes from a Hidden City*, p. 34.

20. Woolfson, *Field Notes from a Hidden City*, pp. 93–4.

Chapter 7: The Wild and the Farmed

1. Jared Diamond, *Guns, Germs and Steel: A Short History of Everybody for the Last 13 000 Years* (London: Vintage, [1997] 2005), pp. 104–13, 309–10.

2. Jay Ferreira, 'Fish Farming: Affordable Protein for SA?' *Farmer's Weekly* (21 September 2013), http://farmersweekly.co.za/article.aspx?id+45208&h+Fish-farming:-affordable-protein-for-SA (accessed 29 September 2014), p. 3.

3. Mike Burgess, 'Pioneering Dusky Kob Production in SA', *Farmer's Weekly* (22 July 2013), https://www.farmersweekly.co.za/agri-business/agribusinesses/pioneering-dusky-kob-production-in-sa/ (accessed 12 September 2018), p. 3.

4. Kobus van der Merwe and Jac de Villiers, *Strandveldfood: A West Coast Odyssey* (Cape Town: Sunbird, 2014), p. 16.

5. Van der Merwe and De Villiers, *Strandveldfood*, p. 14.

6. Van der Merwe and De Villiers, *Strandveldfood*, p. 19.

7. Van der Merwe and De Villiers, *Strandveldfood*, p. 19.

8. It is a manuscript by Frances Cope, the grandmother of Jack Cope. The front pages, and hence the title, are missing, so Martin has simply titled it *Koppie's Story*, after the main character.

9. Lannice Snyman and Anne Klarie, *Free from the Sea: The South African Seafood Cookbook* (Cape Town: Don Nelson, [1979] 1980); Lannice Snyman and Anne Klarie, *More from the Sea* (Cape Town: Don Nelson, [1986] 1990).

10. Van der Merwe and De Villiers, *Strandveldfood*, p. 19.

11. Van der Merwe and De Villiers, *Strandveldfood*, p. 175.

12. Van der Merwe and De Villiers, *Strandveldfood*, p. 19.

13. A term that 'specifically references the speciality limestone and granite Fynbos region that stretches from Postberg in the West Coast National Park, past Langebaan, Saldanha, Jacobsbaai and on toward Stompneus Bay, ending at St Helena Bay'. See Van der Merwe and De Villiers, *Strandveldfood*, p. 14.

14. Van der Merwe and De Villiers, *Strandveldfood*, p. 14.

15. Van der Merwe and De Villiers, *Strandveldfood*, p. 16.

16. Van der Merwe and De Villiers, *Strandveldfood*, p. 16.

17. Van der Merwe and De Villiers, *Strandveldfood*, p. 184.

18. Van der Merwe and De Villiers, *Strandveldfood*, p. 19.

19. Van der Merwe and De Villiers, *Strandveldfood*, p. 19.

20. Van der Merwe and De Villiers, *Strandveldfood*, p. 20.

21. I am grateful to Stephen Boshoff for this insight.

22. Robin Shulman, *Eat the City: A Tale of the Fishers, Foragers, Butchers, Farmers, Poultry Minders, Sugar Refiners, Cane Cutters, Beekeepers, Winemakers, and Brewers who Built New York* (New York: Broadway, 2012), pp. 3–4.

23. Shulman, *Eat the City*, p. 4.

24. Shulman, *Eat the City*, p. 7.
25. Shulman, *Eat the City*, p. 7.
26. Shulman, *Eat the City*, p. 6.
27. Shulman, *Eat the City*, p. 6.
28. Shulman, *Eat the City*, p. 8.
29. Shulman, *Eat the City*, pp. 258–62.
30. Shulman, *Eat the City*, p. 38.
31. Shulman, *Eat the City*, pp. 15–16.
32. Shulman, *Eat the City*, p. 45.
33. Shulman, *Eat the City*, p. 23.
34. Shulman, *Eat the City*, p. 66.
35. Shulman, *Eat the City*, p. 80.
36. Shulman, *Eat the City*, p. 84.
37. Shulman, *Eat the City*, p. 98.
38. Shulman, *Eat the City*, pp. 106, 98.
39. Shulman, *Eat the City*, p. 136.
40. Shulman, *Eat the City*, pp. 98–9.
41. Shulman, *Eat the City*, p. 135.
42. Shulman, *Eat the City*, p. 221.
43. Shulman, *Eat the City*, p. 251.
44. Shulman, *Eat the City*, pp. 221–2.
45. Shulman, *Eat the City*, p. 222.

Chapter 8: Wild Fish

1. The spread of scorpion fish into the Atlantic Ocean through their release from aquaria, and also the global spread of Mediterranean mussels on the hulls of ships indicate that human intervention can cause population growths in marine species, though both of these were inadvertent and undesirable.
2. Conversation with Alan Hobson, who is responsible for stocking the waters in this area, and who has pioneered Somerset East as a premier fly-fishing destination in South Africa.
3. Dave Walker (ed.), *A Guide to Fly Fishing in the Eastern Cape Highlands* (Rhodes: Wild Trout Association, 2017), p. 9.
4. Bob Crass, *Trout in South Africa* (Johannesburg: Macmillan, 1986), p. 62; Walker, *A Guide to Fly Fishing*, p. 9.
5. Paul Schullery, *Royal Coachman: Adventures in the Fly Fisher's World* (Albuquerque: University of New Mexico Press, [2000] 2007), p. 161.
6. Schullery, *Royal Coachman*, p. 161.

7. Schullery, *Royal Coachman*, p. 161.

8. Schullery, *Royal Coachman*, p. 161.

9. Schullery, *Royal Coachman*, p. 162.

10. Schullery, *Royal Coachman*, p. 162.

11. Schullery, *Royal Coachman*, p. 162.

12. Schullery, *Royal Coachman*, pp. 162, 163.

13. Schullery, *Royal Coachman*, p. 163.

14. Schullery, *Royal Coachman*, p. 163.

15. Schullery, *Royal Coachman*, pp. 163–4.

16. Schullery, *Royal Coachman*, p. 164.

17. Schullery, *Royal Coachman*, p. 167.

18. Schullery, *Royal Coachman*, p. 168.

19. Snyder, *The Practice of the Wild*, p. 144.

20. Schullery, *Royal Coachman*, p. 165.

21. Schullery, *Royal Coachman*, p. 170.

22. A lake 'characterised by a low accumulation of dissolved nutrient salts, supporting but a sparse growth of algae and other organisms, and having a high oxygen content owing to the low organic content' (www.dictionary.com).

23. Crass, *Trout in South Africa*, p. 96.

24. Tom Sutcliffe, *Yet More Sweet Days: A Life in Fly Fishing* (Cape Town: Burnet Media, 2019).

Chapter 9: Wild Ethics

1. Jeff Gordon (ed.), *101 Kruger Tales: Extraordinary Stories from Ordinary Visitors to the Kruger National Park* (Leadwood Publishing, 2014), pp. 106–7.

2. Gordon, *101 Kruger Tales*, p. 113.

3. Gordon, *101 Kruger Tales*, pp. 127–8.

4. Livingstone's words are: 'Some poet has a hawk roosting with an inner soliloquy going on about being the monarch of all he surveys [. . .] it may make a pretty picture, it may be supremely clever deployment of language, but it's untrue. I mean it's fair to assume that a hawk, most of his life, is his stomach: spotting, hunting and the rending and tearing of flesh.' See Douglas Livingstone and Tony Morphet, 'Douglas Livingstone in Conversation with Tony Morphet', *Theoria: A Journal of Social and Political Theory* 65 (October 1985): 20–1.

5. Douglas Livingstone, *The Anvil's Undertone* (Johannesburg: Ad Donker, 1978), pp. 18–19.

6. Douglas Livingstone, *A Littoral Zone* (Cape Town: Carrefour Press, 1991), p. 37.

7. Livingstone, *A Littoral Zone*, p. 55.

8. She made this claim in a workshop at which she delivered an early version of the piece that would later appear as a journal article, 'Apartheid's Wolves: Political Animals and Animal Politics' in 2016. It appears, however, to have been excised from the final version.

9. Peter Godwin, quoted in Louise Green, 'Apartheid's Wolves: Political Animals and Animal Politics', *Critical African Studies* 8(2) (2016), Special Issue on 'Writing Animals into African History': 153.

10. Wendy Woodward, *The Animal Gaze: Animal Subjectivities in Southern African Narratives* (Johannesburg: Wits University Press, 2008), p. 1.

11. Woodward, *The Animal Gaze*, p. 1.

12. Quoted in Esther Woolfson, *Field Notes from a Hidden City: An Urban Nature Diary* (London: Granta, 2014), p. 216.

13. Woodward, *The Animal Gaze*, p. 2, quoting Singer.

14. Woodward, *The Animal Gaze*, p. 2, quoting Singer; emphasis added.

15. Woodward, *The Animal Gaze*, p. 3.

16. Woodward, *The Animal Gaze*, p. 5, quoting Plumwood.

17. Duncan Brown, *Voicing the Text: South African Oral Poetry and Performance* (Cape Town: Oxford University Press, 1998); Brown, *To Speak of this Land*.

18. Woodward, *The Animal Gaze*, pp. 4–5.

19. Woodward, *The Animal Gaze*, p. 122.

20. Woodward, *The Animal Gaze*, p. 15.

21. Woodward, *The Animal Gaze*, pp. 5–6.

22. Woodward, *The Animal Gaze*, p. 5.

23. Woodward, *The Animal Gaze*, p. 5.

24. 'Our DNA is 99% the same as the person next to us', https://www.businessinsider.com/comparing-genetic-similarity-between-humans-and-other-things-2016-5?IR=T (accessed 25 July 2018).

25. Woolfson, *Field Notes from a Hidden City*, p. 116.

26. Woolfson, *Field Notes from a Hidden City*, p. 75.

27. Woolfson, *Field Notes from a Hidden City*, pp. 75–6.

28. Woolfson, *Field Notes from a Hidden City*, p. 76.

29. Woolfson, *Field Notes from a Hidden City*, p. 76.

30. Woolfson, *Field Notes from a Hidden City*, p. 149.

31. M. Shelley, (ed.), *The Works of Percy Bysshe Shelley* (London: Edward Moxon, 1847), p. 33.

32. Matthew Scully, *Dominion: The Power of Man, the Suffering of Animals, and the Call to Mercy* (London: Souvenir Press, 2011).

33. Julienne du Toit, 'From Veld to Fork', *South African Country Life* (July 2015), p. 109.

34. Harold F. Blaisdell, *The Philosophical Fisherman: Reflections on Why We Fish* (New York: Skyhorse Publishing [1969] 2015), p. 22.

35. Jonathan Balcombe, *What a Fish Knows: The Inner Lives of Our Underwater Cousins* (New York: Scientific American and Farrar, Straus and Giroux, 2016).

Chapter 10: A Wilder Mind

1. Henry David Thoreau, *Walden* (London: Penguin, [1854] 2016), p. 7.

2. Thoreau, *Walden*, p. 13.

Select Bibliography

Abbott, Andrea. 'Return of the Wild', *South African Country Life*, June 2017, pp. 40–5.

Aplet, Gregory H. 'On the Nature of Wildness: Exploring What Wilderness Really Protects', 1998–1999, http://heinonline.org/HOL/LandingPage?handle=hein. journals/denlr76&div=19&id=page= (accessed 12 September 2011).

Archibald, Brett. *Alone: The Search for Brett Archibald*. Johannesburg: Jacana, 2016.

Balcombe, Jonathan. *What a Fish Knows: The Inner Lives of Our Underwater Cousins*. New York: Scientific American and Farrar, Straus and Giroux, 2016.

Bissolati, Abby-Gene. 'Luiperd "Hou Vakansie" in Huis in Gordonsbaai', *Die Burger*, 10 October 2016, pp. 1, 9.

Blaisdell, Harold F. *The Philosophical Fisherman: Reflections on Why We Fish*. New York: Skyhorse Publishing, [1969] 2015.

Bleek, W.H.I. and L.C. Lloyd. *Specimens of Bushman Folklore*. London: George Allen and Company, 1911.

Brandt, Femke and Marja Spierenburg. 'Game Fences in the Karoo: Reconfiguring Spatial and Social Relations'. *Journal of Contemporary African Studies* 32(2), 2014: 220–37.

Brigg, Peter. 'The River King of Thendela', *South African Country Life*, July 2015, pp. 82–5.

Brown, Duncan. *Voicing the Text: South African Oral Poetry and Performance*. Cape Town: Oxford University Press, 1998.

———. *To Speak of this Land: Identity and Belonging in South Africa and Beyond*. Pietermaritzburg: University of KwaZulu-Natal Press, 2006.

———. *Are Trout South African? Stories of Fish, People and Places*. Johannesburg: Picador Africa, 2013.

Burgess, Mike. 'Pioneering Dusky Kob Production in SA', *Farmer's Weekly*, 22 July 2013, https://www.farmersweekly.co.za/agri-business/agribusinesses/pioneering-dusky-kob-production-in-sa/ (accessed 12 September 2018).

Carruthers, Jane. *The Kruger National Park: A Social and Political History.* Pietermaritzburg: University of Natal Press, 1995.

Cloete, Elsie. '"There's a Meat down There": An Essay on English and the Environment in Africa'. In *Vernacular Worlds, Cosmopolitan Imagination*, edited by Stephanos Stephanides and Stavros Karayanni. Leiden and Boston: Brill and Rodopi, 2015, pp. 21–37.

Comaroff, Jean. *Body of Power, Spirit of Resistance: The Culture and History of a South African People.* Chicago: University of Chicago Press, 1985.

Conradie, Ernst. 'Do Only Humans Sin? In Conversation with Frans de Waal'. Paper delivered at a conference of the European Society for the Study of Science and Theology on the theme 'Are We Special? Science and Theology on Human Uniqueness', Warsaw, 2016.

Crass, Bob. *Trout in South Africa.* Johannesburg: Macmillan, 1986.

De Kock, Leon. 'The Land and its Appropriation by "English"'. In *Literature, Nature and the Land: Ethics and Aesthetics of the Environment* (Collected AUETSA Papers 1992), edited by Nigel Bell and Meg Cowper-Lewis. Ngoye: University of Zululand, 1993, pp. 207–14.

Diamond, Jared. *Guns, Germs and Steel: A Short History of Everybody for the Last 13 000 Years.* London: Vintage, [1997] 2005.

Du Toit, Julienne. 'From Veld to Fork', *South African Country Life*, July 2015, pp. 108–15.

Emerson, Sarah. 'Boston is Covered in Goose Poop and People are as Mad as Hell', 2016, http://motherboard.vice.com/read/boston-is-covered-in-goose-poop-and-people-are-mad-as-hell (accessed 28 October 2016).

Ferreira, Jay. 'Fish Farming: Affordable Protein for SA?' *Farmer's Weekly*, 21 September 2013, http://farmersweekly.co.za/article.aspx?id=45208&h+Fish-farming:-affordable-protein-for-SA (accessed 29 September 2014).

Foster, Craig and Damon Foster. *The Great Dance: A Hunter's Story.* Documentary film, 2000.

Gordon, Jeff (ed.). *101 Kruger Tales: Extraordinary Stories from Ordinary Visitors to the Kruger National Park.* Cape Town: Leadwood Publishing, 2014.

Green, Louise. 'Apartheid's Wolves: Political Animals and Animal Politics'. *Critical African Studies* (Special Issue on 'Writing Animals into African History') 8(2), 2016: 146–60.

Gutteridge, Lee and Louis Liebenberg. *Mammals of Southern Africa and their Tracks and Signs.* Johannesburg: Jacana, 2013.

Haig-Brown, Roderick. *Fisherman's Summer.* Second edition. Ontario: Totem Books, [1959] 1975.

———. *Return to the River: The Classic Story of the Chinook Run and of the Men who Fish It.* New York: Lyons and Burford, [1946] 1997.

Halverson, Anders. *An Entirely Synthetic Fish*. New Haven, CT: Yale University Press, 2011.

Hughes, Ted. *Selected Poems 1957–1981*. London: Faber and Faber, [1982] 1984.

Ingold, Tim. *The Perception of the Environment: Essays in Livelihood, Dwelling and Skill*. London and New York: Routledge, 2000.

Landres, Peter B., Mark W. Brunson, Linda Merigliano, Charisse Sydoriak and Steve Morton. 'Naturalness and Wildness: The Dilemma and Irony of Managing Wilderness'. *USDA Forest Service Proceedings* (RMPRS-P-Vol-5), 2000, pp. 377–81.

Livingstone, Douglas. *The Anvil's Undertone*. Johannesburg: Ad Donker, 1978.

———. *A Littoral Zone*. Cape Town: Carrefour Press, 1991.

Livingstone, Douglas and Tony Morphet. 'Douglas Livingstone in Conversation with Tony Morphet'. *Theoria: A Journal of Social and Political Theory* 65 (October), 1985: 15–25.

Macfarlane, Robert. *Landmarks*. London: Penguin, 2015.

———. 'The Word-Hoard: Robert Macfarlane on Rewilding Our Language of Landscape', 2015, www.theguardian.com/books/2015/feb/27/robert-macfarlane-word-hoard-rewilding-landscape (accessed 27 February 2015).

Mashego, Mojalefa and Gill Gifford. 'Max the Gorilla's Shooter's Riddle', *Independent Online*, 4 November 2005, http://www.iol.co.za/news/south-africa/max-the-gorilla-shooters-death-riddle-257949 (accessed 16 January 2017).

Moffett, Helen. 'A Tale of Two Cities'. *The Big Issue* 237 (23 November 2015 – 5 January 2016): 25.

Monbiot, George. *Feral: Rewilding the Land, Sea and Human Life*. London: Penguin, [2013] 2014.

National Geographic and BBC. *Salmon: Running the Gauntlet*. Documentary film, 2015.

Ndebele, Njabulo S. *Fine Lines from the Box: Further Thoughts about Our Country*. Cape Town: Umuzi, 2007.

Nustad, Knut G. *Creating Africas: Struggles over Nature, Conservation and Land*. London: Hurst, 2015.

Pinnock, Don. *Wild as It Gets: Wanderings of a Bemused Naturalist*. Cape Town: Tafelberg, 2016.

Poynton, Howard. 'The Smell of Rain: How CSIRO Invented a New Word', *The Conversation*, 31 March 2015, http://theconversation.com/the-smell-of-the-rain-how-csiro-invented-a-new-word-39231 (accessed 23 January 2017).

Ramsay, Scott. '9 of the Wildest Experiences in Southern Africa', *Getaway* 26(10), January 2015, pp. 46–57.

Ridder, Ben. 'The Naturalness versus Wildness Debate: Ambiguity, Inconsistency, and Unattainable Objectivity'. *Restoration Ecology* 15(1) (March), 2007: 8–12.

Robinson, Chris and Eileen Finlayson. *Scottish Weather*. Edinburgh: Black and White, 2008.

Rolston III, Holmes. *Environmental Ethics: Duties and Values in the Natural World*. Philadelphia: Temple University Press, 1988.

Roodt, Betsie. *Betsie Rood's 101 Traditional South African Recipes*. Cape Town: Tafelberg, [1977] 1980.

Rose-Innes, Henrietta. *Green Lion*. Cape Town: Umuzi, 2015.

Sandburg, Carl. *The Complete Poems of Carl Sandburg*. San Diego, New York and London: Harcourt, 1970.

Schilthuizen, Menno. *Darwin Comes to Town: How the Urban Jungle Drives Evolution*. London: Quercus, 2018.

Schullery, Paul. *Royal Coachman: Adventures in the Fly Fisher's World*. Albuquerque: University of New Mexico Press, [2000] 2007.

Scott Fitzgerald, F. *The Great Gatsby*. Harmondsworth: Penguin, [1926] 1981.

Scully, Matthew. *Dominion: The Power of Man, the Suffering of Animals, and the Call to Mercy*. London: Souvenir Press, 2011.

Shelley, M. (ed.). *The Works of Percy Bysshe Shelley*. London: Edward Moxon, 1847.

Shepherd, R.H.W. *Lovedale South Africa: The Story of a Century, 1841–1941*. Lovedale: Lovedale Press, 1940.

Shulman, Robin. *Eat the City: A Tale of the Fishers, Foragers, Butchers, Farmers, Poultry Minders, Sugar Refiners, Cane Cutters, Beekeepers, Winemakers, and Brewers who Built New York*. New York: Broadway, 2012.

Snyder, Gary. *The Practice of the Wild*. Berkeley: Counterpoint, 1990.

Snyder, Gary and Jim Harrison. *The Etiquette of Freedom and the Practice of the Wild*. Edited by Paul Ebenkamp. Berkeley: Counterpoint, 2010.

Snyman, Lannice and Anne Klarie. *Free from the Sea: The South African Seafood Cookbook*. Cape Town: Don Nelson, [1979] 1980.

———. *More from the Sea*. Cape Town: Don Nelson, [1986] 1990.

Spierenburg, Marja and Shirley Brooks. 'Private Game Farming and its Social Consequences in Post-Apartheid South Africa: Contestations over Wildlife, Property and Agrarian Futures'. *Journal of Contemporary African Studies* 32(2), 2014: 151–72.

Sutcliffe, Tom. *Yet More Sweet Days: A Life in Fly Fishing*. Cape Town: Burnet Media, 2019.

Thoreau, Henry David. *Walden*. London: Penguin, [1854] 2016.

Van Breda and Others v Jacobs and Others, 1921 AD 330, Appellate Division, Bloemfontein, 15 March 1921.

Van der Merwe, Kobus and Jac de Villiers. *Strandveldfood: A West Coast Odyssey*. Cape Town: Sunbird, 2014.

Versfeld, Martin. *Food for Thought: A Philosopher's Cookbook.* Cape Town: Tafelberg, 1983.

Walker, Dave (ed.). *A Guide to Fly Fishing in the Eastern Cape Highlands.* Rhodes: Wild Trout Association, 2017.

Wolmer, William. *From Wilderness Vision to Farm Invasions: Conservation and Development in Zimbabwe's South-east Lowveld.* Oxford: James Currey, 2007.

Woodward, Wendy. *The Animal Gaze: Animal Subjectivities in Southern African Narratives.* Johannesburg: Wits University Press, 2008.

Woolfson, Esther. *Field Notes from a Hidden City: An Urban Nature Diary.* London: Granta, [2013] 2014.

Index